The Meriden Flint Glass Company

THE MERIDEN FLINT GLASS COMPANY

AN ABUNDANCE OF GLASS

DIANE TOBIN

THE History PRESS

Published by The History Press
Charleston, SC 29403
www.historypress.net

Cover images: Back top: Meriden Britannia Company factory complex, Meriden, Connecticut, late 1800s. This is the firm that built and operated the Meriden Flint Glass Company. Back bottom: Tumbler, Henry S. Fillebrown, United States, Cambridge, Massachusetts, 1890–99, colorless glass; blown, engraved H. 9.1cm. Collection of the Corning Museum of Glass, Corning, New York (2010.4.157).

First published 2012

Manufactured in the United States

ISBN 978.1.60949.492.6

Library of Congress Cataloging-in-Publication Data
Tobin, Diane (Diane Pawloski)
The Meriden Flint Glass Company : an abundance of glass / Diane Tobin.
p. cm.
Includes bibliographical references.
ISBN 978-1-60949-492-6
1. Meriden Flint Glass Company. 2. Glass manufacture--Connecticut--History. I. Title.
TP853.C6T63 2012
666'.1009746--dc23
2012000577

Notice: The information in this book is true and complete to the best of our knowledge. It is offered without guarantee on the part of the author or The History Press. The author and The History Press disclaim all liability in connection with the use of this book.

Dedicated to
Lionel V. DeRagon
and
my mother, Victoria K. Pawloski-Reed

Contents

CONTENTS

Prologue

Here are some interesting events of the 1880s and the Gay Nineties, courtesy of a resident—Russell White, from Meriden, Connecticut, writing in the 1920s—who lived at the North End during those years:

> *Near where the Napier Company's factory is now located, at the eastern end of Cambridge Street, there was, in the 1890s, the ruins of the Meriden Flint Glass Works. This concern was very busy during its short span of life and manufactured table and ornamental glass in many different colors. The glass works plant was destroyed by fire but for many years the tall chimney remained standing until it was struck by lightning, after which it was blown up as it was considered a hazard.*

Introduction

Glassmaking has existed in this country since the ill-fated Jamestown colonists established a glass factory in 1608. That venture failed, but over the next 150 years, the industry continued to develop. By the time of the American Revolution, a primitive glassmaking industry had been established in the colonies.

The chief ingredient of glass is silica, a kind of sand that is mainly fine quartz. In 1662, English glassmaker George Ravenscroft added flint particles to the silica, producing a highly refined type of glass. Recognized for its high clarity and brilliance, flint glass became the standard of excellence in glassmaking.

In the 1800s, it was discovered that substituting lead for flint produced a far superior glass. The lead added heft and clarity, making glass less brittle and able to be cut without shattering. Although now properly called "leaded glass," the earlier term "flint glass" remained in common use as a synonym for fine, high-quality glass.

During the last quarter of the nineteenth century, the American public developed a taste for decorative glassware, and the industry responded with brilliant new colors and innovative designs in cut crystal. In the hands of talented artisans, glass became a medium that produced finished objects of great beauty.

The heyday of American decorative glass production was from 1876 to 1917. During the first dozen years of that span, the Meriden Flint Glass Company operated in Connecticut. Built originally to supply quality glass to the Meriden Britannia Company, the world's largest silver company, it would become the source of glass for the numerous other silver and ornamental metalware companies in central Connecticut that incorporated glassware into their product designs.

Some of the industry's most talented craftsmen worked at the Meriden Flint Glass Company, producing award-winning cut and decorated glass. Names such as James D. Bergen, Thomas B. Clark, Henry S. Fillebrown, Philip Handel, Henry Leighton, Charles F. Monroe, James and Thomas Niland and others could be found on the company's payroll.

The Meriden Flint Glass Company, however, has been largely overlooked in the annals of glassmaking history. Most of its product has been incorrectly attributed to other firms for one reason or another. Its true legacy may be in the influence it exerted on the course of Meriden's development through the introduction of a new industry that would take root in the town and continue to grow long after the factory was shuttered—for which the Silver City would earn a proud second, although not secondary, reputation.

Chapter 1

1876

In the 1870s, the town of Meriden, Connecticut, was home to a group of prominent silver companies that manufactured sterling and silver-plated decorative wares for the sophisticated Victorian home. These firms, responding to a national fad for ornamental glass, began incorporating large quantities of cut and artistic glassware into their product lines to set off their ornate metal goods. Colorful and elaborate glass shades softened the glare cast by kerosene and oil lamps. Fancy glass powder jars and perfume bottles were set into intricate silver baskets for display on dressers. Cut crystal decanters were meant to impress in their sterling cradles, and sparkling glass cruets in argent condiment caddies made passing the mustard an elegant dinnertime event.

Before 1876, all of this glass had to be imported, at great cost, from out of state or even Europe, as there were no glassmaking facilities in Connecticut. The world's largest silver firm, the Meriden Britannia Company, had been dealing largely with the world's largest glassmaker—the New England Glass Company of Cambridge, Massachusetts—as well as importing fine, and costly, Italian glass. When Meriden mayor Horace C. Wilcox, president of the Meriden Britannia Company, toured Europe in the early 1870s, he visited various glassworks on the continent. After closely observing things there, he decided that the same glass could be made in Meriden. Convinced of this, he returned home determined to explore the idea further. He knew that the patronage of his mighty Meriden Britannia Company alone would ensure the success of a local glassmaking operation.

And so it came to pass that, at a special meeting on Monday, January 24, 1876, the directors of the Meriden Britannia Company voted to subscribe for $50,000 worth of stock in a proposed glassworks to be located in Meriden. The factory would be called the Meriden Flint Glass Company. As majority stockholders, the Britannia Company administrators would act as directors of the glassworks.

Four days later, New England Glass Company superintendent Joseph Bourne wrote the following letter to his friend Stephen McCabe, a glassmaker at the Boston and Sandwich Glass Company. The two were old friends and had been corresponding for years.

East Cambridge, Massachusetts
Jan. 28th 1876
Friend McCabe,

Your letter I received some days since but could not answer because for the past week or 10 days I have been in the middle of confusion, if I may so express it. I suppose you have heard by this time that I have left the New England, also the Head Salesman, Mr. Hatch, and we could not be allowed to give in our resignation without giving offence. It was a surprise to everybody, but the matter was a secret. Some of them here call it a conspiracy. However, it is a fact, and the directors decided that Mr. Hatch go forthwith. He left last Friday evening, but Mr. Libbey endeavoured to prevail on me to reconsider the matter and gave me until Monday to decide. On Monday I decided to go, and he then told me the decision of the directors. Mr. Libbey and I parted on the most friendly terms, he felt very sorry of the step I had taken, believing I had been drawn into an affair by those who were plotting to injure the New England. He attaches no blame to me, but feels that Hatch & Wilcox are the great conspirators.

We are going to build a large 10 pot furnace at Meriden, Conn. The president of the Meriden Britannia Company is at the head and everything is fixed. The plans are being made and I am there this week, shall return on Saturday. It will be in a most beautiful part of Meriden. Mr. Wilcox is Mayor of the city and the Meriden [Britannia] *Company will take all the goods we can make. It is a great place for schools, and as nice a city I have ever seen.*

I cannot inform you of much more at present, & I trust you will remain silent upon a few little matters I wish to state to you.

I want good men to make first class work. Can you help me in this matter? Friend, will you come & bring a good shop? You shall have a good position. And is there any others at Sandwich I can approach? I would like to correspond with you upon these matters, as I want a good class of goods manufactured, & the New England & Sandwich I know will show fight. In fact it has already begun. But there is plenty of money and force in the new place to meet them.

Hope to hear from you soon. With my best wishes, I remain your friend,

J. Bourne

The new glassworks was to be located in the north end of Meriden in a section known as Fraryville. William Worcester Lyman and Eli Butler were partners in a land development company that was building up and marketing the area. Mr. Lyman, a prominent local businessman, former city councilor and a member of the House of Representatives, lived on nearby Britannia Street. He was to be the president of the new glass factory.

The site on which the glassworks was to be built was actually a fetid swamp that was reputed to have no bottom. Crows, eels, mosquitoes, water snakes and mussels populated the dank mire, from which a fog arose nightly and where malaria prevailed "to an alarming extent." The road across the marsh had to be built up every spring. Hundreds of loads of dirt were emptied on it, but it continued to sink every year. Horses were sometimes swallowed up and "lost to sight," and men had to "poke around some time" to find them.

The neighborhood would later be marketed as a delightful place to live, although malaria was not completely eradicated for many years, and the unwholesome, swampy odor was eventually replaced by horrific fumes from the glassworks when it started experimenting with crude petroleum in its furnace.

The nation, meanwhile, still recovering from the assassination of Abraham Lincoln and the end of the Civil War only eleven years earlier, was celebrating its 100th birthday with a grand party. Nicknamed the "Mighty Cosmos," the Centennial Exposition opened on May 10, 1876, in Fairmount Park, Philadelphia. Tourists from all parts of the country flocked to the fair on the recently built Transcontinental Railroad. Over the next six months, more than 8 million visitors would gaze at a variety of innovative wonders, including the Westinghouse air brake, Edison's duplex telegraph, the typewriter and the sewing machine. They would marvel at the first electric lights, while a strange little talking device sat unnoticed for weeks on a table near a stairwell. It would be nearly two more years before the telephone would begin to come into trade use with the world's first two commercial exchanges located in New Haven and Meriden, Connecticut.

A glassworks built on site by the Gillinder Glass Company was to become one of the most popular exhibits at the fair. Here visitors could observe the actual process of glass being made and then crafted into sparkling keepsakes. These solidified silica souvenirs, brought back to homes all over the country, would spark a national enthusiasm for cut and artistic glass.

On the same day that the "Mighty Cosmos" opened, frontiersman "Wild Bill" Hickok was shot dead while playing poker at a saloon in Deadwood, South Dakota, and ground was broken for Meriden's new glassworks. Horace Wilcox had arranged with the New York, New Haven and Hartford Railroad to lay a spur track from the mainline to the glassworks. On May 15, he wrote a letter to the president of the railroad company noting that "the building and maintaining of the fences along the line of the proposed track shall be no expense to your railroad company whatever. This I will see is provided and paid for."

Meanwhile, Mr. Bourne continued to correspond with his friend Stephen McCabe at Boston & Sandwich, detailing the construction of the new glassworks at Meriden:

> *August 2, 1876*
> *West Meriden*
> *Friend McCabe,*
>
> *Your letter I received a few days ago. You must excuse my not writing before this time. When I went to Portland I hurt my side a little, and the very hot weather brought out a small rash. When at Boston I did not notice it much, but after I arrived here it commenced to smart & pain me. I went to a Doctor, & he painted it all over with Croton Oil, and took all the skin off which made it very much worse. For several nights I could not sleep or lie down, and could not write to any one. I am improving very much now. I thank you for the sizes of* [lamp] *shades sent—also the model, which I have received.*
>
> *The walls of the Glasshouse are up to the roof and the furnace is up to the top of the arches where they start the cone from. The potarch is being crowned in. The foundation for lears is laid. The storage building is up two stories. The Boiler has arrived & the engine is expected next week.*
>
> *We are commencing to clean out & build some cottages for the men to live in near the works.*

H.C. Wilcox letter to the president of the New York, New Haven and Hartford Railroad Company, May 15, 1876, regarding the building of a spur track to the glass factory. *Author's collection.*

Do you burn oil or rosin or coal in your Glory Holes? And could you find out where the Company get their Fluor Spar and Feld Spar from?

Hope to hear from you soon, and with Best Wishes I remain,
Your friend,

J. Bourne

P.S. I enclose you a view of Gillinders Glass Works on the Centennial Grounds, Philadelphia.

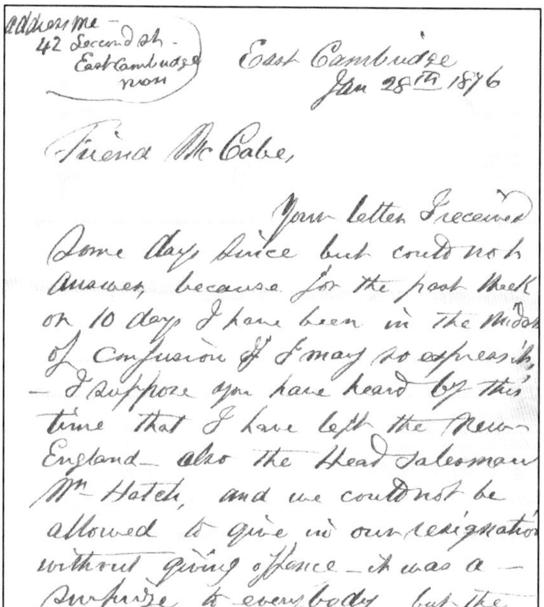

address me - 42 Second st. - East Cambridge now

East Cambridge
Jan 28th 1876

Friend McCabe,

Your letter I received some days since but could not answer, because for the past week or 10 days I have been in the Midst of Confusion if I may so express it, - I suppose you have heard by this time that I have left the New England - also the Head Salesman Mr Hatch, and we could not be allowed to give in our resignation without giving offence - it was a surprise to everybody, but the Matter was a Secret, some of them here call it a conspiracy, However it is a fact, and the Directors decided that Mr Hatch go forthwith, he left last friday evening, but Mr Libbey endeavoured to prevail on me to reconsider the Matter and gave me untill Monday

Above: Envelope addressed to Stephen McCabe, January 1876, at Boston & Sandwich Glass Company. *Meriden Historical Society (MHS).*

Left and opposite: Joseph Bourne's letter to his friend Stephen McCabe, telling him about the plans to open a flint glass factory in Meriden, Connecticut. January 1876. *MHS.*

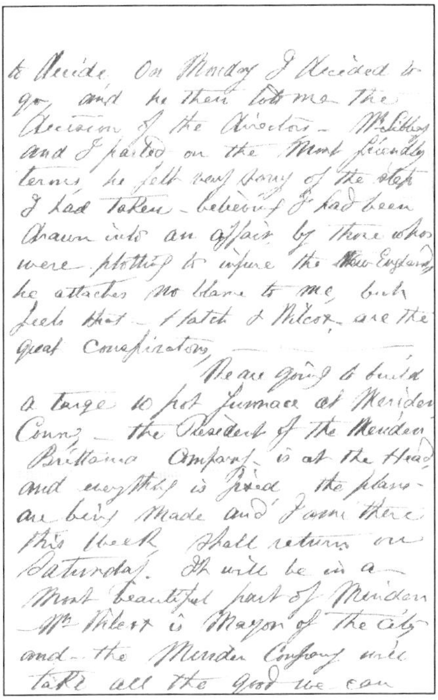

to decide. On Monday I decided to go, and he then told me the decision of the directors — Mr Libbey and I parted on the most friendly terms, he felt very sorry of the step I had taken believing I had been drawn into an affair by those who were plotting to injure the New England, he attaches no blame to me, but feels that Hatch & Wilcox are the great conspirators —
We are going to build a large 10 pot furnace at Meriden Conn — the President of the Meriden Brittania Company is at the Head and everything is fixed the plans are being made and I am there this week shall return on Saturday. It will be in a most beautiful part of Meriden — Mr Wilcox is Mayor of the City and the Meriden Company will take all the goods we can

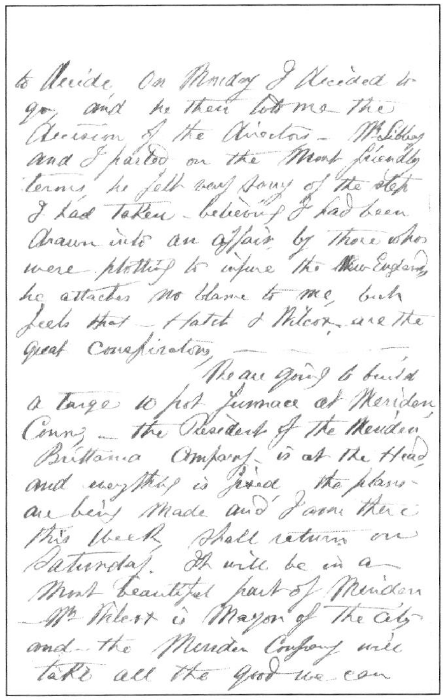

to decide. On Monday I decided to go, and he then told me the decision of the directors — Mr Libbey and I parted on the most friendly terms, he felt very sorry of the step I had taken believing I had been drawn into an affair by those who were plotting to injure the New England, he attaches no blame to me, but feels that Hatch & Wilcox are the great conspirators —
We are going to build a large 10 pot furnace at Meriden Conn — the President of the Meriden Brittania Company is at the Head and everything is fixed the plans are being made and I am there this week shall return on Saturday. It will be in a most beautiful part of Meriden — Mr Wilcox is Mayor of the City and the Meriden Company will take all the goods we can

The local newspaper, the *Meriden Daily Republican*, followed the construction of the glass factory with an eager captivation, reporting regularly on its progress, such as here on August 12, 1876:

> *The work at the glass shop is now being rapidly pushed forward. The building in which the furnace is being built is called the "glass house." This building is nearly done, wanting only an iron roof to complete the outside. The west brick building is already two stories high and will soon be finished. An excellent well has been dug which will furnish an abundant supply of water for the boiler and other purposes. A large number of visitors come and go daily, which shows that great interest is taken in this enterprise.*
>
> *…A broad avenue has been laid out from Griswold street, running into the woods east of the buildings* [Cambridge St.], *which will be immediately put in proper shape. A road, running parallel to Griswold street,* [Tremont Street] *has been staked out in the grove. It is the plan, we believe, to have the workingmen build their homes on this street, thus forming a pleasant community among themselves. Those building lots are so finely situated that other individuals besides the workmen, will, no doubt, desire a spot in this locality.*

By early November 1876, the new glass factory was completed. The furnace fires were lit on the eleventh day of November. This task was performed by Mrs. Henry Warren, daughter of company president W.W. Lyman.

It would take one month for the furnace to reach full blast. It had to attain and sustain a temperature of at least two thousand degrees Fahrenheit without cracking or crumbling before it would be deemed ready for use. Once the furnace reached its full working temperature, the fires would rarely be allowed to go out, as the contraction associated with cooling would be damaging to the structure, causing it to crack.

The Meriden Flint Glass Company began to blow glass on December 10, producing whimsies and items for a local fair. On December 14, 1876, it began manufacturing goods for market. The *Daily Republican* reporter who covered the story seems to have been quite enthused by what he saw. It was customary for factory workers to name their furnace chimneys after volcanoes. This tradition was carried on in Meriden, a fact duly noted by the reporter:

> *Friday, December 15*
> *A VISIT TO THE MERIDEN FLINT GLASS WORKS*
>
> *We paid a visit to the scene of Meriden's new enterprise on Tuesday of the present week, and, under the guidance of the genial manager, Mr. Bourne, made a tour of the building. Mr. Bourne was formerly a manager of the New England Glass Works in East Cambridge, and brings to his new sphere of labor, not only a thorough knowledge of the mysteries of glass making, but also an acquaintance with chemistry, and a culture rarely met with in business circles. It is a genuine pleasure to listen to his explanations, and the hearer cannot help catching some of his enthusiasm and earnestness…*
>
> *The month before our visit, the "Eccles," as the mammoth furnace is not inaptly called, had been fired for the first time, and was now in full blast. The furnace has a shelf around the inside, on which are placed the pots, or crucibles, and each of these has an eye opening outwardly, and situated about two feet above the floor. The pots are manufactured by the company, of the best fire clay, and are so placed upon the shelf as to expose the largest amount of surface to the flames. The mouths, or eyes, are bricked around, and all the observer perceives of them is a round red eye peering forth at the base of the circular brick furnace. The first fire started in the furnace after the pots are placed is an occasion of great anxiety, for if there is any defect,*

they are sure to burst and then they have to be replaced at a great expense of time and labor. But such was the condition of the ten pots, which the furnace contained, that they all stood the test, which is very unusual, and speaks well for the skill and care used in their construction.

This week there were only two shops of hands, as a company of about a dozen men are called, but this week is only the preparatory week, and hereafter, there will be nine shops of men regularly employed. At the time of our visit, the men were mostly engaged in making fancy articles for the Episcopal fair, such as Jacob's ladders, pipes with stems several feet long, etc., articles as light and graceful as frostwork. One of the first-class workmen, Mr. Andrew Long, formerly in the employ of the New England Glass Co., was engaged upon an order of vases for the Wilcox Silver Plate Company, and it was deeply interesting to watch the delicacy and skill with which the fragile articles were fashioned. A long, hollow, iron rod was plunged into an eye of one of the crucibles, and slowly turned, until a sufficient amount of liquid had gathered upon the rod, then the rod

A photograph of a Corliss steam engine in a Meriden factory. (It is not known which factory the picture was taken in.) *Author's collection.*

was withdrawn, the mouth applied to the cold end, and the molten mass expanded slightly, after which it was taken from the rod and while still warm, worked out into the requisite shape. After the article is made, it is put into an annealing oven, and there remains from 15 to 30 hours, according to its size and thickness. There are [four] *of these ovens in this factory, each of which is 67 feet long and open at either end, one into the furnace room and the other into the storage and packing rooms. On the bottom of the ovens, are movable iron plates, connected together, and worked by machinery. On these plates the articles are placed, and in their slow progress through the oven are annealed...*

A side track is laid from the railroad, by which the coal as it arrives, is brought, and emptied from the cars immediately into the engine room. In the rear of the building there are numerous coal sheds adjacent to the track, and a small gasometer, as the company intend manufacturing their own gas...

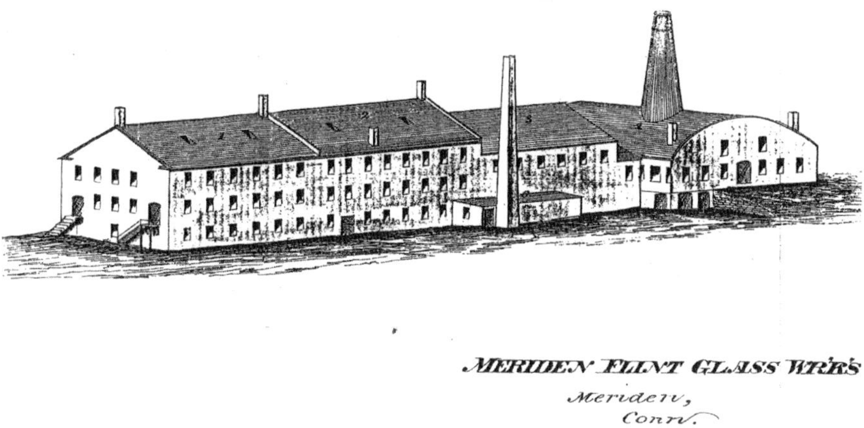

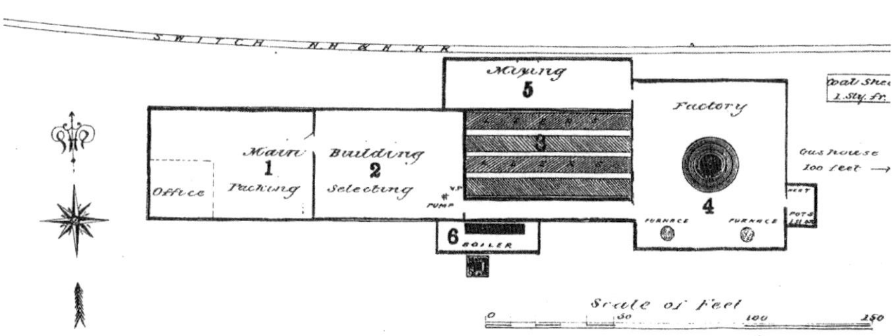

Sanborn Fire Insurance Map of the Meriden Flint Glass Company layout, dated February 1877.

Perhaps the most interesting part of the works is the mixing room, where is seen the various [ingredients] *which are to be transformed into glass. The manner of preparation is something as follows: fine white sand which comes from Berkshire Co., is…dried thoroughly in an oven. After it is dried it is removed and has to be mixed with other substances, for sand alone will not melt. The substances are put in a huge bin, and arranged in layers something as follows. First a layer of sand and then follows layers of red lead* [litharge], *pearl ash, nitre, borax, arsenic, and magnesia, in varying proportions. This mass is thoroughly mixed and sifted, after which it is put into one of the vast crucibles which will contain about 2,000 pounds of glass, where it is melted, the process taking…about* [fifty] *hours, and is then in a crystallized* [clear] *state and is ready to be made up.*

The various colors of glass are produced by addition of other chemical ingredients to the first mixture, as to make blue cobalt is added; for green, iron or copper; yellow, uranium &c.

We have thus endeavored to give some description of a new enterprise, which has arisen in our midst, and which promises to become one of the leading interests in the city. From the character of the men who have undertaken it, there can be no fear about its future prosperity.

Chapter 2

1877

The glassworks' "genial superintendent," Joseph Bourne, was born into a family of glassmakers in 1826 in Stourbridge, England. By age fourteen, he was apprenticed as a glassmaker. He arrived in this country in 1854 and went to work at the New England Glass Company.

At Meriden, he was responsible for the daily operation of the glasshouse, including the hiring and firing of workers, the mixing of all the glass batches, investigating new sources of sand and fuel for the furnaces and much more. He had an intimate knowledge of glassmaking, knew all of his workers by name and was well acquainted with other glassmakers throughout the country. He was concerned for his employees, being quick to "raise a purse" for anyone in need, and he was deeply involved in the community.

Mr. Bourne kept a set of journals in which he recorded daily factory operations in detail, as well as general observations of life outside of the factory. These diaries span the years 1877–78 and 1880–81 and are currently housed at the Meriden Historical Society. Although full of minutiae, the writings are far from mundane. They provide not only a glimpse into the daily functioning of the Meriden Flint Glass Company, as well the lives of the people connected with it, but also, more broadly, a view of the trials and tribulations of running a nineteenth-century glass factory. Newspaper and trade journal articles round out the story.

Joseph Bourne's two journals covering the years 1877–78 and 1880–81. *MHS.*

Diary of Matters Connected with the Meriden Flint Glass Works—Commencing January 1st 1877.

Monday, Jan. 1st
7 pots of glass plain and good colour. White especially good—no pots broken.
Making 7½ in. globes—and no. 3 vases—and vases with thread on top—also mustards & peppers.
The engine not running owing to an unequal balance of the band wheel of [drive] *shaft of cutting shop and other small deficiencies. Sent for an expert from Hartford who understands the Corliss Engine.*
Very much snow on ground to day.

Calling card dishes were popular during Victorian times. These were placed on a table or stand in the entry hall of the home to receive visitors' calling cards. Meriden Flint Glass sales agent George E. Hatch designed both the bird card dish and the inkwell (referred to in the following diary extract) while still employed at the New England Glass Company. The design patents for both were assigned to Mr. Hatch and not to the New England

Glass Company, and so he was able to bring them to Meriden. The bird card dish is pictured in crystal in the Meriden Britannia Company's 1877 catalogue. The inkwells are shown as "Fountain" inks in the same catalogue.

There are no known Meriden Flint Glass catalogues, although the Corning Museum of Glass has two tantalizing pages that appear to be taken from an MFG catalogue. Meriden Britannia Company silver catalogues from 1877–85, however, provide excellent images of the glassworks' products that were made for that company.

Many glass items, such as the inkwell and bird card dish referenced below, were produced in a mold. First, a wood or plaster pattern made by a mold-maker would be sent to the factory's blacksmith shop, where a rough iron cast would be made. The iron mold was then milled and shaped and the design cut into the surface. Iron molds for glassmaking had to be preheated to approximately one thousand degrees Fahrenheit before use, so that the two-thousand-degree molten glass would not crack them when poured in. A gather of liquid glass drawn from a pot would be blown into the prepared iron mold by a glassblower, transferring the pattern to the glass.

> *Thursday Jan. 4th*
> *Making Bird Card Dishes—mold worked well—tried also the ink—worked well.*

Every week, new batches of dry ingredients were mixed, or "made up." This was quite a process in itself, requiring precision and a general working knowledge of glass chemistry. It is described well in the December 1876 newspaper article shown earlier.

Each Friday, the premixed powdered batch materials were brought to the furnace in wooden vessels and shoveled into the mouths of the pots, which were already in place inside the furnace. When all ten pots were full, every opening would be bricked over and sealed with fireclay, and the furnace would be brought up to full heat of two thousand degrees Fahrenheit. The pots remained sealed inside the furnace all day Saturday and Sunday. During this time, small openings would be made, and a little of the metal would be withdrawn to check its progress. This was called proofing, or taking proofs of, the glass. Any depleted pots would be topped off at this time if needed.

Superintendent Bourne recorded in his journals every Friday the number of pots that were being filled, as well as with what type of glass. While most of the workers usually had weekends off, Mr. Bourne himself came in every Saturday and Sunday to check the pots and monitor their progress.

On Monday morning, all was ready for the glassmakers. The pots were full of melted batch. A shop of workers assembled at each pot, and the mouths were opened, providing access to the melted glass inside, which looked like liquid fire. The metal would be worked out by the men during the week and fashioned into predetermined glass items. By Friday, the pots would be replenished with dry batch materials that had been prepared during the week, to be melted into liquid glass over the weekend.

> *Friday, Jan. 5th*
> *An abundance of Glass, coming good.*
> *The men worked 3 moves, making same as previous day.*
> *Filled in* [Pot] *No. 3 x; No. 6, 7 & 9 xx; and No. 4 opal.*

Above: George Hatch's "Fountain Inks" as represented in the 1877 Meriden Britannia Company catalogue. *MHS.*

Opposite inset: "Bird" card dish shown in crystal (flint glass) in the 1877 Meriden Britannia Company catalogue. *MHS.*

Workers from other glasshouses were actively and continually recruited to come and work at the Meriden Flint Glass Company. This was common practice among glasshouses, and the flow of workers often ran in the opposite direction, too!

Meriden Flint Glass Company engraver Henry Leighton had been apprenticed to New England Glass Company First Engraver Louis Vaupel, one of the preeminent glass artisans of the nineteenth century.[1] He came to work at the fledgling Meriden Flint Glass Company, where he would remain for the next ten years. The New England Glass Company, at which Leighton's family had worked for generations, was on the verge of failure. Henry wrote the following in a letter to his father in January 1877: "Yesterday afternoon [New England Glass Company engraver] Cram made his appearance and is going to work here as soon as his machine arrives and is put up. He left the N.E. last Friday and says that last week his earnings amounted to $7. He was the best engraver they had at the time he left and now the most work they have is cheap cruets."[2]

Mr. Bourne recorded the arrival of new workers during this weekend entry, as well as an unfortunate weather mishap:

> *Saturday, Jan. 6th*
> *Topped off the pots, all right, ware coming out good.*
> *3 of cutters came from N.E.* [New England Glass Company] *to day.*
> *In the afternoon went down to Meriden Britannia Works—came up to Glasshouse, and went home—spent the evening there.*
> *This has been a week of good fair working. Glass has come plain, and considerable glass worked out.*
> *During the night the snow fell upon the roof of Boiler house and stove it in.*

The following day, Mr. Bourne noted not only the condition of the glass and what was being made but also the condition of one of his men:

> *Monday, January 8th*
> *No.4 opal & No. 6 xx only plain.*
> *Making dome shades & tall flower glass.*
> *Steven reported not fit for work from wiskey* [sic].

Crucibles, special fireclay pots in which the raw materials for making glass were melted, were built by hand on-site, for which the glassworks employed several full-time pot makers. A completed pot stood about four feet tall, was about three inches thick and over a yard across and held about two thousand pounds of metal (or molten) glass. Pots for melting flint glass had a hoodlike cover to prevent furnace impurities from falling in. The reason for this is that the interior chamber of a flint glass furnace was only about five feet high. The heat and flames reverberated from the ceiling, or crown, of the furnace directly down on the top of and around each pot. For this reason, a flint glass furnace is called a reverberatory furnace.

There was a small opening, or "eye," high up on one side of the pot from which liquid glass could be extracted. When the pot was being set in the main furnace, the eye had to be positioned so that it faced outward and lined up with a small opening in the wall of the furnace so that the glassmaker could insert a blowpipe into the pot and withdraw the molten glass.

A constant supply of hand-built melting pots was kept on hand, to be used as needed. Before being placed into the main furnace, a cold pot had to be tempered to furnace temperature so that it would not crack when exposed to the extreme difference in temperature. A small furnace, called a pot arch, was constructed near the main melting furnace. Cold pots brought up from storage were placed in the pot arch to gradually bring them up to white-hot furnace temperature.

Once set, a pot was never removed from the main furnace until it cracked. It was refilled over and over again. The average lifespan of a working pot that had survived the tempering was from two to three months.

Mr. Bourne here records putting two new pots in the pot arch to ready them for the melting furnace. He also notes the quality of some of the work being produced:

> *Wensday, January 10th*
> *Put 2 Pots in potarch.*
> *Moran's French ring shades—¾ of them condemned—marks around the rings—welt on edge—and glass inside rough.*
> *The Bird dish with acid on opal looked splendid.*

Four years earlier, a benevolent society called the Knights of Honor had been founded. This was a sort of precursor to modern-day life insurance policies. Among the Knights of Honor's principal objects was the social, moral and intellectual development of its members, as well as the payment of death benefits to the widows and orphans of deceased members. Headquartered in St. Louis, it had a grand dictator, grand lodges and numerous subordinate lodges all over the United States. Meriden hosted the Wilson Lodge, as noted in the January 13, 1877 *Meriden Weekly Republican*: "Wilson Lodge, K. of H. of this city, publicly installed their officers at G.A.R. [Grand Army of the Republic] Hall last night. They were installed by Past Dictator G.E. Hatch. After, came readings, rec'd letters, addresses, &c.…and splendid singing by a quartette composed of August and Henry Hirschfield, Otto Henriques and George Lohmann. The party present, though small, had a very enjoyable evening."

Bourne continued in his diary:

> *Tuesday, January 16th*
> *Steven commenced after his sick drunken spell.*
> *Glass all good, and everything doing well.*
>
> *Thursday, January 18th*
> *Birds coming out crisled & black marks.*
> *White not coming good—not melted on the founding properly.*
> *Mr. Dodd of the Silver Plate Co. reported that the Castor Bottles sent were very cordy.*

Coal was an indispensable item in a nineteenth-century glass factory. Before the discovery of western natural gas supplies, virtually all gas was made by heating coal in sealed vessels called retorts. When it became very hot, the coal broke down into three components: gases, tar and coke. The gases so produced gave a brilliant light when burned. This process for converting coal to combustible gases was called coal gasification. The Meriden Flint Glass Company produced its own gas for lighting in retorts and stored it in gas storage tanks called gasometers. Later, gas pipes would be laid to the glassworks from the local city gas company to supply the larger amounts required for the big furnace.

> *Saturday, January 20th*
> *The gas tank* [gasometer] *burst about 10 a.m.*

The city did not yet have a high school, and therefore great interest was demonstrated in improving the lot of working men and their families through lectures. The local newspaper began promoting the idea of a lecture series:

A Lecture Circuit
January 22, 1877

With the various industries we have here in Meriden, all involving scientific skill, there is no reason why we could not have one of the best, or, at least, the most interesting course of lectures, ever presented to our citizens…Let us illustrate: We have just established here a factory for the making of glass ware. We are told that the Meriden Flint Glass company's Superintendent, Mr. Bourne, understands the entire history of the discovery of glass making, that he is a fine fluent speaker, and could, if he chose to do so, give a very interesting lecture on the subject. Why not Mr. Bourne be…urged to prepare and deliver this lecture in Town Hall! He could be followed by other gentlemen in the different factories…Such lectures could hardly fail to attract the attention of working people, and would get them interested in literary pursuits of a practical nature, for the more we hear of a subject, the more we want to know of it, for the human mind is of an inquiring turn, and naturally thirsts for knowledge, under proper cultivation. We bring this matter before our readers for discussion, because we believe it will be of practical benefit to have these lectures, and we hope Mr. Bourne will be urged to open the Course, and tell us "what he knows about glass," which, by the way, would not be a very bad title for his lecture.

A great amount of heat is required to melt glass. A fire was built in the center of the furnace floor in a firebox called the eye, located below the level of the siege. The fuel, usually coal, rested on an iron grate in the eye. Beneath the furnace was the cave, a ten-foot-high underground open-ended air tunnel. A current of air was drawn up from the cave by the draught of the chimney cone, passing up through the iron bars to feed the fire. Flues carried the smoke from the furnace to the chimney.

The fuel was fed from the floor of the glasshouse, shoveled into openings in the furnace wall called tease holes, from which passages sloped down to the firebox.

Meriden's furnace consumed two and a half tons of coal per day. As the price of coal rose, Mr. Bourne began to experiment with alternatives fuels, including different varieties of wood, gas, steam and the latest innovation, petroleum oil.

Friday, February 2nd
Had wheeled in 20 tons of coal.

Glass all right...Gathered out a Ball of Clear Glass—[from Pot] *no. 6—the Best and Clearest Glass—also Colour—made yet—Water Colour.*

Saturday, February 10th
Tried gas in Glory Hole to day—operated well—a little improvement needed.
Furnace not going at regular speed—alternated bye holes a little at 5 p.m.

In addition to problems with alcoholism, there were occasional employee run-ins with the law.

Wensday, February 14th
Today a detective from Sandwich came and enquired for Montague and arrested him for having been concerned in a robbery at Sandwich of a clothing store.

A few days later, the Meriden newspaper reported:

We are informed that Michael E. Montague, the man who was arrested here last Wednesday...on suspicion of being concerned in a burglary committed in Sandwich, Mass., and taken to that place to be tried, has been released, as nothing could be proved against him, and the articles found in his trunk could not be identified. Montague is back again at work at the Meriden Flint Glass works, and has a poor opinion of Massachusetts, as besides losing several days' time, he had to pay his way back to Meriden. Massachusetts detectives seem to think that because one criminal was traced to Meriden, that every man who commits a crime in that State, flees immediately to this city.

Nitrous oxide, or "laughing gas," originally used in sideshow entertainment, had first been used as an anesthetic by a Hartford dentist thirty years earlier.

Monday, February 19th
In the fore noon went down town at 10 a.m. and had two teeth extracted to the gas which made me feel very queer through the day.

Tuesday, February 20th
Got a severe cold to day—& feeling very unwell.

Made to day the first Hero Shade by attaching the flint [clear] *to the white by joining the two edges together to give a reflection of light downwards.*

Wensday, February 21st

The white & flint shade a success. Will show light with either kerosene or gas, and requires no chimney.

I have a very severe cold which gives me pain all through my body—went home—put my feet in hot water and went to bed.

Thursday, February 22nd

Feel a little better to day but very miserable.

Mr. Wilcox and all parties much pleased with the shade. Commenced to make a large number.

Had a little fuss with [worker] *Gorman.*

Saturday, March 3rd

Pay day. Proofs all right. Topped off the pots 8 a.m.

Men & shops docked for seconds. Chase 460 globes, 50 cts.;

Chase very mad and intends to send a globe to Sandwich.

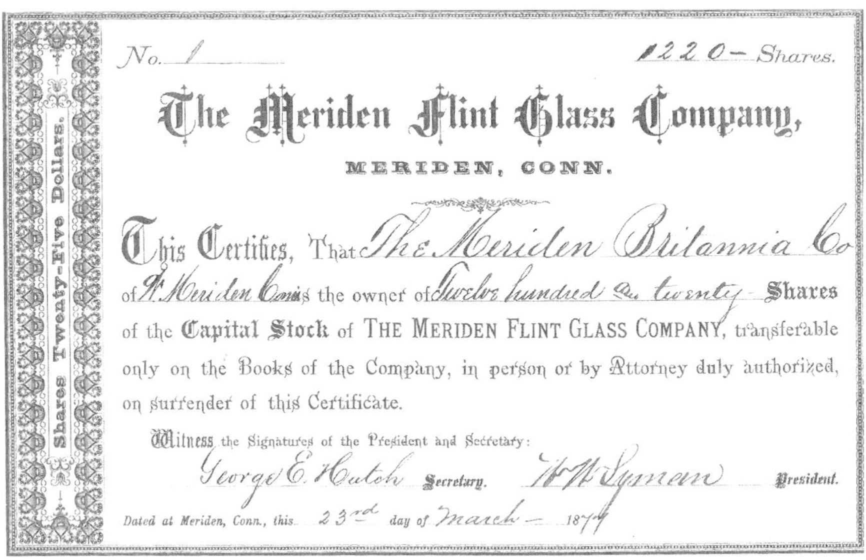

No. 1 1220— Shares.

The Meriden Flint Glass Company,

MERIDEN, CONN.

This Certifies, That The Meriden Britannia Co of W. Meriden Conn is the owner of Twelve hundred & twenty Shares of the Capital Stock of THE MERIDEN FLINT GLASS COMPANY, transferable only on the Books of the Company, in person or by Attorney duly authorized, on surrender of this Certificate.

Witness the Signatures of the President and Secretary:

George E. Hatch Secretary. H. W. Lyman President.

Dated at Meriden, Conn., this 23rd day of March 1874

Shares Twenty-Five Dollars.

Meriden Flint Glass Company stock certificate for 1,220 shares assigned to the Meriden Britannia Company. *MHS.*

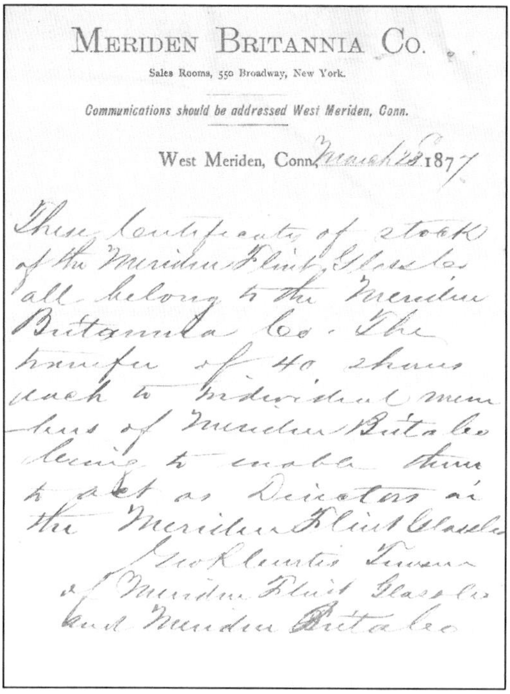
MERIDEN BRITANNIA CO.

Sales Rooms, 550 Broadway, New York.

Communications should be addressed West Meriden, Conn.

West Meriden, Conn, March 28 1877

These Certificates of stock of the Meriden Flint Glass Co all belong to the Meriden Britannia Co. The transfer of 40 shares each to individual members of Meriden Brita Co being to enable them to act as Directors in the Meriden Flint Glass Co

Geo R Curtis Treasurer of Meriden Flint Glass Co and Meriden Brita Co

Letter dividing the Meriden Britannia Company's shares of MFG stock among company administrators, enabling them to act as directors of the glassworks for the silver company. *MHS.*

Monday, March 5th
President Hayes inaugerated [sic] *to day.*
Glory hole running fair with wood. Put Gorman on the lears to day. Whitherbee down in Sloar.

Newly made glass had to be slowly cooled, or annealed, so that it would not crack. There were four lears, or annealing ovens, for this purpose at the Meriden Flint Glass Company. Each low-vaulted brick oven was nearly seventy feet long and ran from the glasshouse to the Sloar Room in the next building. New glassware was placed in iron pans at the glasshouse end of the lear. The pans were hooked together and mechanically drawn slowly through the heated tunnel. There was a fire at the glasshouse end of each lear, and as the articles passed along the tunnel, moving farther from the flames, they slowly cooled, arriving at room temperature into the Sloar Room. Here the pans of annealed glassware were removed from the lears and sent to the decorating and cutting departments. The empty pans were sent back to the glasshouse to be hooked together and filled again.

Above: Exterior view of the south end of the brick lear building, showing three of the four lear tunnel openings (one is boarded up). *CEI Investment Corp.*

Left: Inside one of the four lear tunnels, 2011. *CEI Investment Corp.*

Tuesday, March 6th
Nothing of importance. Mr. Lyman in New York. I delivered a lecture on glass in G.A.R. Hall—well attended.

Mr. Lyman was at attorney Earle's office in New York because he had patented a design for a glass alumet, or match-holder.

On March 7, 1877, the local paper reported the following:

The lecture given by Mr. Bourne was largely attended, and among the audience were many of the most prominent people in the city. The tickets were only obtainable from members of the Knights of Honor, and they were in great demand during the day. The lecture was both exhaustive and interesting, and a great deal of original information was given. The history was taken up from the earliest time at which glass was mentioned by ancient writers, and its progress traced down through the ages until the present time. Many interesting facts and stories were given, which enlivened the more matter-of-fact portions. The interest of the audience was well sustained, even to the end, and, although the lecture was a little more than two hours in duration, yet there was nothing in it which could well have been omitted. The large attendance and general interest show that our citizens are interested in gaining useful information, and we wish that the lecture could be repeated in a larger hall, and thrown open to all classes of citizens, as it is an opportunity for benefiting the general public, which should not be let pass.

During the evening, the Cosmopolitan club gave some very fine songs.

Bourne returned to his diary.

Saturday, March 10th, 1877
Mr. Hatch sent a letter stating that he had an order for 1,000 doz. Hero Shades, and that Mitchell, Vance & Co. were pleased with our ware.

When a crucible cracked inside the furnace, molten glass flowed from the pot down through the furnace grates and into the cave below, where it cooled and solidified. This waste glass, called cullet, was periodically collected and recycled as a batch ingredient, where it acted as a flux and prevented the formation of clumps. Pots and their contents were expensive, so recycling was essential. Not only was the waste glass harvested, but the broken pot shards were also ground and mixed with raw clay to make new pots.

Above: Mitchell Vance and Company catalogue, Plate No. 7, showing Meriden Flint Glass Company decorated opal ring-top cone shades. *Paul Ivazes Quality Lighting.*

Below: Mitchell Vance and Company catalogue page showing Meriden Flint Glass etched frosted lampshades. *Paul Ivazes Quality Lighting.*

Pot-setting was an arduous process. First, the firebricks were pulled away from the front of the furnace without reducing the firing, so that the flames flared out in all directions. The furnace heat had to be maintained so as not to damage the other pots. A gang of men approached the furnace with a kind of "chariot" consisting of two long rods on wheels. The prongs of this "pot-trolley" were pushed under the cracked pot, and the men pressed down on their end to lift the white-hot pot from the floor of the furnace. They then drew away with the glowing pot balanced on the prongs and wheeled it away to a place where it could be tipped and emptied of any remaining contents. They then moved to the pot arch and repeated the process to pick up the new, equally glowing pot, which they trundled into the furnace and set onto the siege. The pot arch was located very near to the main furnace, so that the heated crucible could be quickly transferred to it without cooling off. The men then picked up the firebricks on long thin shovels and placed them into position to cover the front of the furnace. The bricks were then coated with wet fireclay to seal the gaps.

Mr. Bourne recorded the first broken pot with but little fanfare:

> *Sunday, March 11*
> *½ past 7, the side of* [Pot] *No. 6 broke out and let the glass into the cave.*
> *Glass in* [Pots] *No. 7 XX, 8 XX, 9 XX & No. 1 opal plain. Topped off No. 3 opal.*
> *Let the furnace down.*

The furnace temperature was raised very high during the weekend in order to melt the glass and then was let down for Monday so that the men would be able to approach the furnace to work the glass.

On Tuesday, March 13, 1877, Mr. Bourne received a letter from his friend Stephen McCabe, asking for work. McCabe reported also that Gillinder's had "made a miserable attempt to make" Hero Shades years ago. Further, Mr. Hatch heard that the Mount Washington Glass Works was making a Hero Shade and was attempting to prevent Meriden from obtaining a patent for it. It was decided to send Stephen McCabe there to investigate.

Cost Control

Lime glass was a lower grade of glass ideally suited for common tableware. Glasshouses that were sprouting up near midwestern coal fields produced great quantities of this cheaper lime glass. When abundant supplies of natural gas were discovered in the Midwest, some midwestern towns offered free natural gas to any company willing to relocate there. The combination of cheap fuels and the mass production of lime glass was to pack a powerful punch to the East Coast glass industry. East Coast glass factories still had to import their fuel at great expense, which canceled out any profit associated with the mass production of glass. For this reason, East Coast glasshouses continued to rely on producing quality glass by skilled artisans.

In the next journal entry, we find that Mr. Wilcox wanted to investigate cheaper production methods for the glass cruets, which were made in vast quantities for the Meriden Britannia Company's silver-plated dinner castors. He also wanted cruet stoppers to be made in a lapidary mold, to look as if they were hand-cut and faceted. He was trying to lend an air of distinction to a mass-produced item without increasing its production cost.

Monday, April 2nd
10 in dishes made last week, many came out cracked from the blow over.
Mr. Wilcox here to day and it was decided not to make any more castor bottles at present, as we were putting the very best glass into them at a low figure.
Mr. Wilcox desired me not to make any more stoppers in new mold. He would have a lapidary one. He also stated that they were making bottles from lime glass—castor bottles—at 50 cts. per dozen.

The *Daily Republican* reported on April 3, 1877, that "in a large iron factory in western Pennsylvania all the heat for smelting is made by gas, which is brought in pipes from a natural well nineteen miles away."

The glassworks occasionally bought pots from an outside manufacturer.

Thursday, April 5th
Received a letter from Coffin [pot manufacturers] *stating that 14 pots were shipped April 2nd, seven of them having rings in them, and seven without. Sizes of the Pots—36 x 48 x 42.*

The local paper reported on April 11, 1877, that "six additional new cottages are going up near the Glass Works. Several nice homes are being erected, which will make this quite a cheerful village."

Amenities in the "cheerful village" included an arrangement for the delivery of meat to the cottages at a discounted price, with the cost being deducted from the men's pay. A ten-pin alley had been constructed for the amusement of the neighborhood. There was also a baseball field in Gehrmann's Grove behind the glass factory. By June, it was said that "the Meriden Flint Glass Co. has the strongest baseball nine in the city."

Since coming to town, Mr. Bourne and his corps of glass men did much to promote the growth of temperance and other reforms. The underlined names in the following newspaper article from May 7, 1877, were all glassworks employees:

> *Town Hall was nearly filled at the temperance meeting Sunday afternoon. Mr. Bourne presided and opened the meeting with some earnest remarks upon the subject of temperance. Short speeches were also made by Messrs. Stannis, Cram, Cochrane, Curtis, and others. Mr. Hamlin led the singing,*

An Artizan Street "cottage" in 2011, built by the Meriden Flint Glass Company for glassworkers' families in 1877. "Artizan" is the nineteenth-century spelling of "artisan."

Another Artizan Street workers' cottage. Five of the original residences built on this street in 1877 are still standing and occupied in 2011.

View of the glassworks from Artizan Street today.

in which the audience heartily joined. A temperance prayer meeting will be held at the Methodist church this evening, to which all are invited, and there will be the usual Wednesday evening meeting at the Old Road school-house, under the leadership of Mr. B.P. Ray.

Daily exposure to toxins such as lead and arsenic used in the making of glass often resulted in chemical poisoning among glassmakers. Mr. Bourne was no exception. He, as well as many of his employees, was often taken sick for extended periods of time.

Mr. Bourne described these episodes as characterized by chills or shakes, with shortness of breath and a feeling of great weakness. He was often unable to work during these spells.

As there was no hospital in Meriden at this time, people suffering from injuries or illness were tended at their homes by local physicians who made house calls. The doctor was generally not called for unless the situation seemed dire.

The first episode of Mr. Bourne's sickness is recorded here. These were to occur with alarming frequency over the next five years. Mr. Bourne often tried to work through his sick spells.

Friday, May 11th
Afternoon taken very sick, left the [pot] *filling to Beach.*

Saturday, May 12th—sick.

Sunday, May 13th—sick.

Tuesday & Wensday, May 15th and 16th—sick.
On Thursday came around a little, made up two batches.

Saturday, May 19th
Feeling better to day, but weak.

Saturday, June 2nd
Whalen the Teaser hurt his hand very much.
Topped the pots off and filled in No. 10 with opal.
This forenoon I was taken with the shakes and went home sick.
Put Cady on through the remainder of the founding.

Monday, June 4th
Glass opened fair. No. 6 XB dull colour. White very good.
Taken with another shake at 12 p.m. which was a very violent one, and I was very sick through the day.

Tuesday, June 5th
At the Works and coloured a Batch, went home.

Wensday, June 6th
At home all day.

The Meriden Flint Glass Company supplied glass tubes for a local thermometer-maker named Mr. Kendal. The following is a description of blowing thermometer tubes, or tube-drawing. The operation depends on the fact that when a hollow lump of glass, such as a globe or a cylinder, is drawn out while hot, the interior cavity remains intact, even if the glass is drawn out as thin as fine thread.[3]

First, a blower inserted a blowpipe into a pot, withdrew a gather of clear glass and blew gently into it. He would then flatten the glass slightly on a cast-iron marver, flattening the large bubble of air that he had just blown into it as well. At this point, the bubble was about as broad as a finger and extended the full length of the glass. The elongated bubble would become the bore of the thermometer, containing the mercury and the white backing against which it would be visible.

Next a gather of molten opaque white glass was touched to the back of the now-hardened clear glass and drawn from end to end along the flat side, leaving a white stripe nearly as wide as the bubble within.

The clear glass, with the white strip pressed down and hardened on it, was now plunged into and thickly coated with molten clear glass. The soft exterior was rounded on the marver and then reheated until soft again.

At this point, the blower adhered the melting lump to the end of a punty, and a boy ran with it in one direction while the blower backed off in the other, the molten glass stretching between them like gum. When they were seventy or eighty feet apart, the gleaming cord of glass sagging between them was laid on the ground. The bore and white stripe had stretched with the glass tube, while keeping their flattened shape.

The tube was then cut into sections about five feet long, packed into crates and sent off to the thermometer-making shop. There the sections were cut down into tubes of the right length. One end of each tube was melted, sealed shut and blown out into a small bulb. To avoid introducing any moisture into the tube, a small rubber bag was used to puff air into the open end, inflating the far end into a small bulb.

The small tube was then heated, causing it to expand and expel the air from the bore. The resulting partial vacuum drew the mercury up from the bulb end and into the bore. To expel any remaining air, the mercury was heated, expanding to as high a temperature as it was expected to measure, and then the open end was heated and sealed shut. The mercury within contracted as it cooled, leaving a vacuum at the upper end of the tube, into which it could expand as heated.[4]

Tuesday, June 19th
Commenced to draw thermometre tube for Mr. Kendal, Moran's shop drew 130 lbs. 2 moves.

Some days at the factory were better than others. The following is not one of them.

Monday, June 25th
White coming fair. No. 4 XB very speky. Sand stones.
Tried the new sand in monkey. Could not melt it. Added more flux, pot broke, all run out.
Running small hole with coal—run fair. Wood better.
Very warm day.

The Meriden glassworks depended on trains for delivery of all the raw materials used in the making of its glass, vast amounts of coal to fuel its furnace and the transport of its finished goods to retailers all over the country. Railroad strikes disrupted supply deliveries to factories, which delayed its production of goods for market and resulted in lost profits.

In July, the Baltimore and Ohio Railroad reduced employee wages by 60 percent, resulting in rail strikes in five states, according to the local paper on July 20, 1877:

The Great Strike
It Is Spreading to Other Roads

Although there is not such an exhibition in favor of violence, the railroad strikers on the Baltimore and Ohio road still stand out, and are being joined by employees of other roads, just as predicted on Thursday. Four trains were started from the depots Thursday afternoon, but, in order to protect them, they had to carry military on the freight cars. As a coal train moved off several strikers occupied the hill along the side of the track, which was covered with rocks. The military presented arms, and there was no demonstration.

Considerable drunkenness is noticeable among the rioters, although the liquor stores are closed.

Monday, July 23rd
Rioting in Pitsburgh & other places south & west.

Tuesday, July 24th,
The mobs of Philadelphia, Chicago, St. Louis, Pitsburgh still on the fighting and destroying path. All transportation west & south of New York suspended.

And, of course, if it wasn't one thing, it was another:

Wensday, July 25th
Crown of Potarch sagged in on the pots, burning in the afternoon. Got the pots out safely and put them into 9 & 6 arches, and cooled the Pot-arch down. Sent for the men who were not working & who came to help.

It would take a full three weeks to repair the pot arch.

The Meriden Flint Glass Company's massive furnace consumed two and a half tons of coal per day. As a result of the rail strike, the price of coal began to rise dramatically, from twenty-five cents to fifty cents per ton, as noted in the July 30, 1877 *Meriden Daily Republican*. By August 1, the rail strike had ended, but "the coal dealers generally are having quite a lively sales on account of the recent rise in the article and apprehension that the prices are still in the ascendant," noted the local paper on August 6, 1877.

Monday, August 6th
Pot-arch all finished to day. Took out the center and put in a little fire to smoke it.
Men working all night.

Tuesday, August 14th
My birthday, 51 years of age!

Put on the draft to pots in Potarch through the night Monday night. Arch sweating and swelling out. Kept pots at ½ draft through the day. Put on full draft through night.

Wensday, August 15th
Keeping the pot arch on half draft, pots look well, and the arch runs well.

"The local coal dealers advanced half a dollar on a ton today," noted the paper on August 16, 1877. The next day saw this: "Freight Receipts—Flint Glass Co. 26 cars coal."

Legislators wrestled with the issue of child labor in the nineteenth century. Although factory working conditions could be abysmal, there were no social services or poverty relief programs in existence then. Preventing children from working often reduced their families to the level of paupers. Children were paid dramatically less than adult workers, though, and many factories relied heavily on their cheap labor in order to remain in operation.

In the extreme heat of the glasshouse, there were children as young as twelve years of age working around the clock alongside the men, and they were sometimes the targets of abuse. Mr. Bourne fired or had arrested anyone who struck or otherwise harmed the boys in his factory. He noted several such instances in his journals.

Friday, August 17th
At 6 p.m., O. Hamlin struck Cady's Boy with a hot piece of glass, making a large mark on his face. Boy reported he had been throwing at him during the afternoon. Cady very mad, I advised him to get out a summons.

Most Connecticut factory workers outside of New Haven County were paid once per month.[5] Within New Haven County, which included Meriden, more than two-thirds of factory employees received their paychecks on a weekly basis.

Many employers felt that workers who were paid monthly often ran out of money and had to resort to purchasing food and necessities on credit, forcing them to shop only at those stores where credit was issued. They argued that employees who received weekly pay always had cash and could shop anywhere they could get the best value for their money.

Another issue that employers took into consideration was alcoholism, which plagued many factories. The consequences of a "night out" were most apparent the day after payday. Many employers preferred to pay their workers on a monthly basis, if only to cut down on the frequency of employee drinking binges, hangovers and "sick days."

Many employers also felt that having to pay employees every week or biweekly represented an increase in work for their bookkeeping staffs. This was another reason for them to continue with the monthly payroll tradition.

Saturday, August 20th
Pay day, the day men & teasers starting on the same plan of paying as glassmakers, one week in hand.

It was industry practice to shut down glass factories during the heat of the summer, as the men could not tolerate working in both the natural and artificial heat. These shutdowns usually lasted for a number of weeks, during which time the furnace would be dismantled and examined for damages. The Meriden Flint Glass furnace had only been running for a few months, however, and was still "like new," needing no repairs. On August 24, Mr. Bourne shut the factory down anyway, due to the heat, and went to New York to visit various glassworks there, as well as the dealers who were selling Meriden Flint Glass products.

By August 31, Mr. Bourne had returned to Meriden, "not feeling well at all."

Saturday, Sep 1st
Took of proofs, fair.
Came home to dinner, felt very unwell and at one o'clock had a chill & fever which lasted until 5 p.m. Very sick in the evening and through the night.

Sunday September 2nd
Went to glasshouse—felt weak. Everything right.

Monday, September 3rd
Glass plain & good. No. 9 opal broke in bottom. Worked out the glass, set the men working and about 10 a.m. took the shakes and had a very severe one, made me very sick indeed.

Tuesday, September 4th
Very sick during the day, unable to go out.

Wensday, September 5th
At home through the day. Franklin around glasshouse. Feeling better to day.

Thursday, September 6th
Went to glasshouse and around during the day, but feeling very weak.
Andy Long drawing tube.

Every year since 1871, a combination agricultural/arts and crafts/manufacturing fair was held at the town hall, acting as a showcase for local industry. Manufactured goods and poultry were on view at the town hall auditorium, while livestock were exhibited at the trotting park off Bailey Avenue in the north end of town. The September 18, 1977 local paper reported:

The Fair!
Magnificent Opening
A Gala Day at the Park
Town Hall Presents a Grand Appearance

The Meriden Flint Glass Co. is a new feature, and the display of decorated shades, ornaments, vases, and a line of rich cut glass-ware will not fail to attract the admiring crowds, who will applaud the men whose energy and enterprise have been instrumental in establishing this industry in Meriden....

Bradley & Hubbard during the afternoon had erected in the hall, and will have lighted to night, the magnificent 24-light gasalier that attracted so much attention at the Centennial. It is a splendid looking affair, and will help to add to the beauty of the scene this evening. The chandelier will be fitted with the Lyman Crown globe, the invention of Hon. W.W. Lyman of this city, and manufactured only by the Meriden Flint Glass Co.

...Mr. F. Miller, at the Meriden Flint Glass Co., shows half a dozen specimens of oil paintings, which are striking in appearance and of much merit. Mr. Miller, evidently, is an artist of no mean order.

Bourne continued in his diary.

Wensday, September 19th
Fair opened at Town Hall and show of cattle &c at Meriden Park. Made some 8½ in. Crown Globes for fair.

Saturday, September 29th
Mr. Hatch…engaged a Mr. Monroe from New Bedford as a Painter.

Tuesday, October 9th
Moran & Hamlin not working.
Hamlin finding fault, considered he was not treated well, &c.

Monday, October 15th
[Farmer] *Clarke gave the men & boys some grapes for their good conduct in not stealing them.*

Thursday, October 18th
Mr. Teall here from Hartford to know about the possibility of making glass coffins.

Wensday, October 24th
Went up to Town Hall with Mr. Hatch. He and I were admitted as voters in the City of Meriden.
Went up to Hartford and gave my opinion on Glass Coffins, which was not very favourable to the Party who had the patent.

Monday, October 29th
A disgracefull affair of drunkenness came to light to day on the occasion of Cady's wife going to Portland. Moore, Wilkinson, Franklin, Ruks, McNaul, King, I. Collins & Cady. The result was in finally discharging King & W. Wilkinson.

A popular toy at the time was a glass curiosity known as a Prince Rupert's Drop. Glassmakers created them by dripping molten glass into a bucket of ice-cold water, creating a teardrop-shaped item with a long, thin tail. The rounded end of a Prince Rupert's Drop is so strong that it can withstand blows by a hammer, yet breaking even the very tip of the thin, fragile tail causes the entire drop to explode into a fine powder!

These were supposedly discovered by Prince Rupert of Rhine in the mid-1600s, although they may have been known before then. The prince's uncle, King Charles I of England, is said to have used the drops as a practical joke. He would give one to an unsuspecting person and then break the tail, causing a small explosion in the surprised victim's hand.

Sunday, November 4th
Everything going on well. Commenced to make Prince Rupert Drops out of No. 4. Made them through the night.
Mr. Wilcox spoke to me to day about articleing for five years.

Monumental Clocks, also called "Apostolic" Clocks, were popular traveling exhibits in many cities. Often standing over ten feet in height, they featured animated carved figurines, astronomical indicators, mechanical organ music, chimes and bells. Dozens of these large and complicated handcrafted clocks toured the United States and Europe. Their purpose was to awe and entertain, and the public flocked to see their intricate displays. Noted the November 16, 1877 paper:

> *The Great Strasburg clock opens Monday its exhibition in Morse & Cook's Block. It is really worth seeing, as an instructive idea of the genius of man. It is a fac simile of the Roman clock at Strasburg, and is known as the Apostolic clock. It was the entire life work of a European peasant, who undertook the wonderful work of copying the Apostolic clock. To see this critical piece of mechanism is really worth three times the admission fee. Tickets are only ten cents each.*

Louise Pomeroy was a popular actress who had been touring in Pennsylvania, New York, Ohio and New England, accompanied by her husband and her own acting company. She had played nearly all of Shakespeare's heroines, as well as the male roles of Hamlet and Romeo. Her company was presenting *Romeo and Juliet* in Meriden, with Miss Pomeroy as Juliet. They performed in the city hall's third-floor auditorium.

Tuesday, November 20th
Miss Louise Pomeroy here this forenoon. Took her around. Had a goblet made for her. Gave me two passes to see her play Juliet at Town Hall in the Evening. Went with [daughter] *Etta.*

Wensday, November 21st
Miss Pomeroy's goblet out. Had it engraved "Louise" in a wreath.

Saturday, November 24th
Went to see the Strasburgh Clock.

Mr. Bourne occasionally capitulated to feelings of annoyance:

> *Thursday, November 29th*
> *Thanksgiving Day. Very wet, miserable and unholy day. Glassmakers working, except a few of the aristocrats. Very unwell myself throughout the day.*

Noted the December 5, 1877 paper: "It is reported on the street that James M. Bartlett is about to start an omnibus line, running from the Corner to the Glass Works. It will leave the Corner near the Meriden House on every hour. Such a line is much needed."

Although the Meriden Flint Glass Company produced its own gas in retorts, it also had gas lines run from the city gasworks to its plant to supply the large amounts of gas required to fire its furnaces. According to the December 8, 1877 paper: "One of the rather remarkable sights on Britannia street is that of a German woman at work in the ditch which is being dug for the gas pipes leading to the Glass Works. She attracts considerable attention from the passers by but takes it all as a matter of course and handles her pick and shovel in as skillful a manner as any of the men. She wants no favors of any man and evidently is able to keep up the best of them."

> *Monday, December 10th*
> *One year to day since we commenced to Blow Glass.*

> *Tuesday, December 25th*
> *Christmas Day. Very beautiful day. No work. Everything running well.*

The December 27, 1877 *Meriden Daily Republican* reported on the omnibus:

> *The new omnibus arrived in town this morning, having been driven up from New Haven by Mr. Bartlett. It is really an elegant affair, and would compare favorably with any of the Broadway stages. It is handsomely painted, has red velvet cushions and the inside is decorated with pictures, and upon the outside is the legend "Colony Street, Depot, Organ and Glass Works." On the lower part of the body is the name of the proprietor, J.M. Bartlett. It is drawn by Mr. Bartlett's well known span of greys, which for size and strength can not be easily matched. New harnesses add to the style of the whole turnout, and Mr. Bartlett says if the whole thing does not come up to any rig of the kind there is going, he*

A drawing of the Meriden Flint Glass Company done by employee C. Robert Fasy circa 1879. Note the two cottages in the background. These were later moved to Tremont Street to make way for railroad tracks. The omnibus in the foreground has "Bartlett's" on its side. *MHS.*

> *will buy another that will. We do not think he will have reason to buy another very soon.*

Bourne maintained his diary:

> *Monday, December 31st*
>
> *And last day of the year. It died surrounded by sunshine, without the snow. A very beautiful day. Glass opened good. All hands working.*
>
> *A very beautiful day. This has been to me a year of sickness and great anxiety, which rendered it a continual drag upon my system, but to day at the close of the year I feel well & upon the whole the result of the year's labor has been fair. There seems to be a satisfaction generally that we have done well under the circumstances.*

CHAPTER 3

1878

Joseph Bourne continued to keep up his diary:

> *January 1st, 1878*
> *Very beautiful clear day.*
> *Great numbers out making calls. Large number of visitors at the Works.*

The January 2, 1878 *Meriden Daily Republican* noted local New Year's Day receptions: "Another very popular place was Mr. W.W. Lyman's residence, where Mrs. Lyman and [daughter] Mrs. Henry Warren welcomed their friends."

The Meriden Flint Glass Company was starting to experience some labor discord. One year earlier, Mr. Bourne had a meeting with his men to discuss inferior work. He informed them that the company would allow one dozen "seconds" on every move before docking their pay. This was agreed to by the workers at the time, but as the number of "seconds" grew over the course of the year, so did tensions.

> *Tuesday, January 29th*
> *Today an order for 10 in. dishes went back in consequence of the cutters refusing to cut them for 50 cts. each. Wanted $1.75 for one kind & $1.50 for the other.*

W.W. Lyman's old house on Britannia Street, reputed to have a covered wagon built into it. The Lymans lived here for thirty-five years before building the "new" house immediately to its right in 1881. Both are still standing in 2011.

Friday, Feb. 1st
Pay Day. J. Moran's shop docked for 2 doz. 12 in. Shades, 4.00. Moran very mad &c.

Saturday, February 16th
Called the men together this forenoon and stated the following reduction to be made, as follows: Proposed that a reduction of 10 P.Ct. be made general, with the exception of the Globe & Shade shops, as regards the Gaffers, who shall receive Pr. Move $2.37½ cts in future, and that 110 good globes will be expected from single shops, and 130 good ones from double shops in future.
The castor Bottle shops will work by the hundred and be paid for whatever is packed, at the rate of 17 cts. Pr. hundred. For Peppers & Mustards and for Vinegars the Blowers to receive 21 cts pr. hd. [per hundred]. *The Gaffers 25 cts. Pr. Hd. Subject to the same conditions as the above, Peppers and Mustards.*
The men appointed a committee to correspond with various places to find out the price of wages and numbers made, and to report on the 23rd. The Castor Bottle makers also to submit a rate of Prices at the same time.
Notified the teasures, mixers & labourers of the 10 p. Ct. reduction.

There would be "much dissatisfaction felt at the reduction of 10 pr. Ct." in the coming months.

On January 28, the world's first commercial telephone exchange opened in nearby New Haven. Serving twenty-one customers, the switchboard, designed by George W. Coy, could connect as many as sixty-four customers. Only two conversations could be handled at the same time, though, and six connections had to be made for each call.

Present at the first public demonstration of Coy's switchboard was Meriden telegraph operator Ellis B. Baker Sr., who had been curious about how to build a telephone exchange. George Coy supplied Baker with the drawings and information that he needed to construct his own.

Armed with Coy's plans, Baker returned to Meriden and had machinist William C. Homan build two telephone receivers. One was installed in Homan's home at 139 Crown Street. It is not known where the other receiver was set up, although one can speculate that it was at Mr. Baker's residence.

Working with Roger D. Blish of the Miller Company, Baker produced an operational switchboard by January 31, 1878, making Meriden, Connecticut, the home of the world's second telephone exchange.[6]

Mounted on a black walnut board, Baker's seven-circuit switchboard included teapot cover knobs made by Meriden's Miller Company. It was installed at A.L. Stevens Real Estate, Insurance and Steamship Company at 31 Colony Street.

Brass wires strung between wooden poles and attached to porcelain knobs connected the phones.

Within one month, there were forty-seven subscribers to the new service. Among these was the Meriden Flint Glass Company, making it one of the world's earliest commercial telephone customers.

The original switchboard operators were boys who had been trained as telegraph operators. They were soon replaced by women, who were found to be more courteous.

The New Haven switchboard was later destroyed in a fire, leaving Meriden's the oldest switchboard in the world. It is preserved at AT&T's History Center in Warren, New Jersey.

The *Rigg's Meriden Literary Recorder* of March 2, 1878, reported that "the construction of the Meriden district telephone lines will be commenced this week. A telephone wire is to be constructed to run from the Britannia works to Manning & Bowman's, Wilcox Silver Plate company's, Glass Works and H.C. Wilcox's residence."

Meriden's original 1878 switchboard, the oldest one in the world. *AT&T Archives and History Center.*

With the appearance of telephone poles about town, Meriden residents began to experience "telephone fever." The *Daily Republican* noted on March 7, 1878:

> *Dr. Wilson's lecture on the telephone, at the Corner church Wednesday night, was listened to by a large audience, and proved even more interesting than the one which was given on the same subject before the lyceum a few evenings ago. The different points were fully illustrated, and the explanation of the theory and construction of the telephone was remarkably clear. There were some dozen telephones in the church, which were so connected with the wires that they could be passed from hand to hand, and nearly everybody had an opportunity of listening to what the "wild waves" of sound were saying. Wires had been run from the church to Dr. Wilson's house, to the*

office of Dr. Rust, and also to the Western Union office, and had there been connected with New Britain and Hartford. During the evening secular music was sung and played at Dr. Wilson's house, and sacred music at Dr. Rust's office, and the people in the church who listened at the telephones could distinguish the different tunes. Parties in New Britain and Hartford also sang several songs, which could be heard quite as distinctly as the music from the places in this city. The audience were very much delighted with the evening's entertainment, and at having learned something in regard to this wonderful new invention.

Newly made glassware was drawn through the lears for tempering on open metal pans, exposing the wares to flame and smoke from the fire that heated the ovens. This exposure often discolored the glass. The lears were also drafty, causing temperature variations that weakened the glass, especially those items that were later to be cut. In order to overcome these difficulties, the company began experimenting with ways to protect the glass during annealing. It tried placing the glass items in airtight metal boxes instead of on open metal trays. The boxes then passed through the lears on a metal conveyor belt, with the enclosed glass pieces protected from contact with smoke, flame and drafts.

Thursday, Mar. 7
Found most of the bowls put into the boxes melted.

Saturday, Mar. 9
This has been an exciting week. The glass came very good, expecially No. 5 pot, and some nice goods were made.
The men had a meeting to day and a committee appointed who proposed to the company that 5 pr. ct amount be reduced, and no docking. The 5 pr. Ct. could not be entertained, and Mr. Hatch proposed that the men on globes run one month, at the end of that time if all was satisfactory nothing would be done in the matter, if not, something would have to be done.

Monday, Mar. 18
Glass opened well, white & flint.
Trays that were put in boxes in lear came out good & sound.

The patent for the company's "new method of tempering glass" was issued on April 16, 1878.

In April, Mr. Wilcox and Mr. Bourne met with workers Hamlin and Moran and proposed price reductions for globe seconds. Mr. Wilcox also stated that if the price of globes came up, a proportionate rise should be given to them. The men went away, decided not to make them and did not work again through the week.

It was also decided to have all broken and unsalable ware booked every day, and reports were made to the men on a daily basis. Mr. Bourne and bookkeeper William Butler commenced to make up a price book for production costs.

The April 15, 1878 *Daily Republican* noted that President Lincoln had died thirteen years ago on that day.

> *Monday, April 22*
> *Had a very severe head ache this afternoon.*
> *20 Bbls. Adkins Lead arrived today.*

The month of May saw "Speckey" glass, poor work, crooked shades and globes cut wrong. Bad coal made iron slag in the furnace. Some employees went home with the shakes, while others cleared out without paying their board, and entire shops were "discharged" for poor work. Mr. Bourne had shakes, chills and weakness. He continued to work, however, setting pots, mixing batches, taking proofs and making up the price book.

On May 1, 1878, Paris opened its third World's Fair. Running through November 1, 1878, it was the largest exposition ever held anywhere in the world, occupying more than sixty-six acres and attracting 13 million visitors. France showcased the newly completed head of the Statue of Liberty, while the United States displayed Alexander Graham Bell's telephone, Thomas Edison's megaphone, the phonograph and electric light bulbs. U.S. exhibitors could send their goods free of cost by the USS *Constellation.*

There would be extensive glass displays from many nations, including France, Austria and England. The Meriden Flint Glass Company was the only American glasshouse to exhibit. It "somewhat reluctantly" sent a small sample of cut goblets and champagne glasses, one or two cut glass trays and some decorated lampshades, for which the European jury awarded the company two honorable mentions. Mr. Bourne never mentioned it.

The local paper on June 13, 1878, reported that "Meriden manufactures are getting a world wide reputation. The Paris correspondent of the *New York Evening Post*, in a recent letter concerning the Exhibition, gives a very complimentary notice of the really fine work in flint glass shown by the Meriden Flint Glass works."

In mid-June, Mr. Bourne started ordering materials in anticipation of the glassworks' first summer shutdown to examine and rebuild the furnace. Over the next few weeks, he purchased eye bricks, clay and common firebricks. Mr. Libbey from the New England Glass Company stopped by for a visit. Meanwhile, production continued, and before the factory was shut down, a new trade union would be born.

> *Wensday, June 26, 1878*
> *Making thermometer tube for Mr. Wilder.*
> *Trying the infernal octagon bottles today.*
> *Very hot.*

Birth of the American Flint Glass Workers Union

The Meriden Flint Glass Company had been experiencing periodic episodes of labor unrest, generally in response to the company's attempts to control costs by imposing wage reductions for poor-quality work.

Earlier attempts by Pittsburgh and New York/New Jersey–area glassmakers to organize had been unsuccessful. In the 1870s, though, Pittsburgh-area glassworkers began to organize successfully under the auspices of an umbrella group, the Knights of Labor.

In 1878, they made a major effort to form a national union in an attempt to standardize wages and working conditions among factories in the East and Midwest. East Coast glassworkers, including those in Meriden, were invited to a convention held in Pittsburgh on July 1, during which the American Flint Glass Workers Union (AFGWU) was organized. Twelve locals were chartered, with four representing Pittsburgh; two for Steubenville (Ohio); and locals in Brooklyn, Philadelphia, Wheeling (West Virginia), St. Louis, Bellaire (Ohio) and Meriden.[7]

Ironically, the union's activities to equalize wages resulted in the demise of the glassmaking industry on the East Coast by encouraging the removal of its glass factories to the Midwest, where it was simply cheaper to do business.

By their participation in the formation of the AFGWU, the Meriden Flint Glass Company employees were involved in the development of collective bargaining in the United States.

Nationwide collective bargaining was initiated in this country in 1886, when a number of glass manufacturers sent a letter to the president of the AFGWU. The letter stated that the manufacturers had appointed a committee that desired to meet with a like committee from the union to discuss and, if possible, agree on move lists and rules for the ensuing blast.

These committees met in July 1887 and agreed on lists and rules for the 1888 blast. This was the first known instance of collective bargaining not only in the glass industry but also in the United States on a nationwide scale.[8]

Mr. Bourne, notably, made no mention of the union activity in which his workers were involved.

Monday, July 1, 1878
Glass opened good and good colour.
Fogo off—drunk x x x x x x
McCabe making opal fruit baskets.
Extremely hot day.

Bottom of an opal fruit bowl showing the Meriden Britannia logo fired onto the glass. *Author's collection.*

Tuesday, July 2, 1878
Very hot day—but all hands working.
Shipped the 18 in. tray to Collamore, New York.

Wensday, July 3, 1878
Pay day—Fogo's shop left off on account of the heat.
Received 475 fire brick common.
The hottest day yet.

Ten days later, Meriden's "Eccles" furnace was shut down for the first time since November 1876 to be inspected and refurbished. This would become an annual event. It had been operating for over one and a half years, running at temperatures from one thousand to three thousand degrees Fahrenheit nearly the whole time.

The first step was to take down the "front of pots," the ten bricked-up arches behind which the pots were sitting, so they could be removed from the furnace. A team of bricklayers and masons arrived to do the work.

Saturday, July 13, 1878
At one p.m. commenced to take down the front of Pots & let out furnace—all bank'd up, could not take them [the pots] *out except No. 9.*
Drew the bars at 6 p.m., left the pots standing until Monday.

Monday, July 15
Commenced and took out all the Pots—furnace cooling down—3 cracks on top of Crown—over [Pots no.] *10—6—3.*

Tuesday, July 16
Got out nearly the full round of the eye—burnt out 2 feet 3½ inches from the edge of eye.

Wensday, July 17
Progressing on the getting out of siege. Took out the top course—very badly burnt—chipping off top—very much slag on.
I went home in afternoon sick.
Franklin off sick.

Thursday, July 18
Commenced to repair the lears.

Making progress on the siege.
Hottest day yet—102 in city.

Friday, July 19
Laid the first fire brick of foundation of siege at ¼ to 10 a.m.
The new eye plate weighed 2,180 lbs—2 in. thick.

Meanwhile, the following Monday, the *Hartford Daily Courant* reported: "The [Meriden] Flint Glass works make the only American exhibit of glass ware. They show goblets and wine glasses, but have not made an extensive display. It is at least of interest that this young concern should be the only one in the line of manufacture to show anything, especially when glass is said to be an available export article for American makers.—Hf'd Courant's Paris Letter."

Friday, August 2nd
Finished the siege at 5 p.m.

Monday, August 5th
The clay on the top of siege cracked in places—pounded it over and laid sand over the top of siege.
Commenced to build up the arches where pots go.

Tuesday, August 6th
Put in the first fire to day the siege.

Wensday, August 7th
Fire in eye of furnace hot and warming whole thing thoroughly.
Carting away ashes to fill up hole on the south side of Cambridge Street.
Started to New York at 10 p.m. and took the New Haven boat at 11 p.m.

Mr. Bourne visited a number of glassmakers during what was to become an annual summer pilgrimage to New York City, where the company was selling its glassware. He visited dealers such as Mitchell, Vance and Company and the Meriden Britannia showrooms there. He also purchased materials, hired workers and socialized with other glassmakers. He was especially impressed that year during his visit to Tiffany's.

Saturday, August 8
From Mr. Nichols I went to Meriden Britannia store, from there to Tiffanys. Had a long and very interesting conversation with Mr. Haveland, who has charge of the glass department. He had some of the best finished ware I ever saw...His opinion of glass manufactured in this country is that it is not finished well in the cutting shop, not bringing out the brilliancy of the colour.

Arriving back in Meriden at 9:00 p.m. on Friday, he noted in his journal: "This evening the greatest calamity that ever befell the town of Wallingford happened about 6 p.m, caused by rain—lightning—and thunder."

Tuesday, August 13th
Building up the flues, blacking the pillars of furnace, and finish whitewashing the furnace to day. All ready for the fire.

Wensday, August 14th—My 52nd birthday.
Put fire to the furnace at 12 minutes past one o'clock p.m., [son-in-law] *Charles Lake applying the match to the fuel.*
Went to Wallingford to see the place where the tornado made so much destruction.

Thursday, August 15th
McCabe went to Boston to see Mr. Libbey, upon his—Mr. Libbey's—invitation. A. Long reported to day that Mr. Libbey had written to him to come and work for him.

Saturday, August 17th
Andy Long & McCabe informed me they had decided to go to the New England to work.

Monday, August 19th
This morning made an arrangement with Frank Henriot for 12 months to give him not less than 8 moves per week, at two dollars and fifty cents per move. Joseph Ray also promised me he would stay with the Meriden Flint Glass Works.
Sent for McCabe—had a long talk with him, and also went to Mr. Wilcox with him, who also persuaded him to stay with us. He finally consented—upon the condition of receiving <u>Three dollars and twelve cents per move.</u> He then wrote a letter to Mr. Libbey declining to go to the New England.
Andy Long went to Boston to day.

Since August 12, eleven empty new pots (nine full size and two monkeys) had been placed on the siege around the inside wall of the refurbished furnace to temper them as the furnace was brought back up to full heat. The arches were then opened, without letting down the furnace's temperature, and each pot was examined to make sure that they had withstood the heat without cracking. The inside of each pot was then glazed with a thin layer of melted cullet to seal it. The arches were then bricked over, leaving a small opening through which the dry, premixed batch material could be shoveled into the now sealed-in and heated pots.

> *Tuesday, August 27th*
> *Furnace in good condition, had all the pots taken down & examined. All good and sound. Filled in at ½ past 9 a.m. some white cullet to glaze the pots.*

Rigg's Meriden Literary Recorder, on August 31, 1878, noted that "the Meriden Flint Glass Works received honorable mention at the Paris Exposition in Classes XII and XVII."

In response to pleas for help from the yellow fever–stricken South, the mayor requested donations from the citizens of Meriden. The glassworkers decided to raffle off a cut glass bowl during the annual fair and donate the proceeds to the yellow fever fund.

The *Daily Republican*, on September 18, 1878, reported the following:

> *EXHIBITED AT THE FAIR*
>
> *Let us direct your attention to the solid exhibits, of which there are many, and which must, of necessity, be noticed only briefly…Turning around from the organs, the beautiful display of cut glass and decorated lamp shades…by the Meriden Flint Glass Co., greets the eye, and among this grand display is the beautiful glass bowl to be drawn for on Friday evening, the proceeds over the cost of making to go to the yellow fever fund. It might be said that a lady will be in attendance to sell tickets for 50 cents each to those who desire to have a chance in the drawing, and help the sufferers at the same time.*

The glass dish proceeds of $27.25 were donated to Mayor Lines for the yellow fever fund. F.H. Shaw had the winning ticket, No. 72.

Etched flint sugar shaker, with "Wilcox" etched on the bottom. This would have been purchased as a blank and decorated by Wilcox cutters. *Author's collection.*

Tuesday, October 1, 1878
White coming middling—making Belle wines, clarets, champs, 7 in. by 4 globes.
T. Wilkinson went home in the forenoon, sick.
Sick myself in the afternoon, but did not go home.
John McCabe not working—not in a condition for work.

Saturday, October 5
Gave [union representative] *McElheny a resolution to present to the G.M.* [Glass Maker's] *Society—*
Resolved, that any member of this Society losing his time through intemperance shall be held responsible and pay the full value of the time any other member of the society shall lose on his account—and the intemperate member shall also pay the full value of all unsalable goods made through his action of intemperance—the company for which he labors to receive the damages.

Monday, Oct. 14
Fogo in Liquor but at work. J. McCabe not working.
Making bottles and white globes, flower vases, &c.

The October 28, 1878 *Hartford Daily Courant* had this featured article:

Paris—The Meriden Flint Glass Works, which were started after the close of the Centennial exhibition, have sent a small but good sample of their cut-glassware. It consists chiefly of goblets and champagne glasses with one or two trays all deeply cut in the geometrical style, in vogue a century ago and now so generally revived. The cutting is excellent and the form and "metal" good. Great credit is due to the enterprising founders of this important industry at Meriden. They already manufacture largely, and in addition to standard articles for household use, make a great variety of small objects suitable for mounting in silver or white metal by the Meriden Britannia Company. Decorated lamp shades, which are shown here with the cut glass, are also successfully made in large quantities. Although the company was averse to making any exhibition, and did not attempt display commensurate with the present and expected magnitude of the works, the jury have decreed an honorable mention. This, under the circumstances, may be regarded as a high award. At no previous exhibition has there been such an extensive and varied display of glass of all kinds and from many countries. The French and Austrian displays are particularly brilliant. England has also many exhibitors of blown and colored glass.

After the close of the Paris Exposition on November 1, the following report was issued:

Report of the Commissioner-General,
Universal Exposition at Paris, 1878
Our Success at Paris
by Richard McCormick

The Meriden Flint-Glass Works, of West Meriden, Connecticut, sent some fine cut table and decorated glassware—a creditable indication of what must, before long, become an important branch of manufacture in this country.[9]

Bourne maintained his diary.

Tuesday, Nov. 19
Mr. Kendal came about 9 o'clock for thermometer tube—commenced to draw about ¼ to 10 o'clock a.m. Henriot made a poor show in making the bore, put W. Jukes on—made a poor days work.
Glass notty.
Fogo went home drunk.

Wensday, Nov. 20
Drawing thermometre tube, doing well. Everything coming good.
J. McCabe went home with the shakes—or rum fever.

Christmas Day—1878. Very beautiful day. No working.

CHAPTER 4

1879

Rigg's *Meriden Literary Recorder*, on May 17, 1879, was the first local newspaper to hint at trouble brewing at the glassworks. While doing so, it also provided a very informative essay on some of the details of glassmaking:

> *There is talk of closing the works of the Meriden Flint Glass company in this city, for the coming season, on account of the inability to make it pay the cost of operation. These works are the only establishment of the kind in the state, where glass is blown and made into glass ware, ornamental and plain, cut and uncut, is in this city…The glass works are a mile north of the depot, and are reached by an omnibus which makes hourly trips.*
>
> *Visitors are not refused admittance and are entertained by an insight to the many various ways this industry is conducted. To those who have never witnessed glass blowing…the visit is full of wonder and instruction.*
>
> *The blowing, that is the glass making, is done in a large room of which the ground forms the floor. In the centre is an immense furnace, fed with streams of oil, kept at a heat of 3000* [sic] *Fahrenheit, and containing half a dozen or more large clay pots, each having an outlet through windows or openings in the furnace.*
>
> *Every Friday the mixture, which when fused becomes melted glass, is poured into these pots which are made to withstand the fearful heat as the bottoms are kept constantly at a white heat, but become useless after three or four months' service. The principal ingredients which form the fuse are silica, obtained from Berkshire county, Massachusetts, some compound of*

soda, white or red lead, according as translucent or transparent glass is desired, and some other minerals in smaller quantities…

Monday the mixture has become thoroughly fused and is ready for use. Then the work of blowing is carried on day and night until Friday again. Workmen of all ages are employed, the boys of twelve to the experienced workmen of eighty…They must approach close to the furnace to get the supply of melted glass from which the bottle, or goblet or lamp chimney is to be made. But the glass blowers have become accustomed to this sort of thing and appear to suffer but little from the intense heat. With hollow iron rods five or four feet long they pick up out of the pots inside glass from which to blow whatever article they are making…The amount is gauged by turning the rod around and around; the fusion clings to it and is taken out at a white heat…The workman begins blowing through the iron rod the minute the lump is out of the pot. In the case of articles made from opal glass as for instance the shades used with argand burners for which white lead is used, the material is very porous and cools very rapidly and before the desired shape is reached, it must be heated again and again by holding it in the furnace.

The process by which one of these shades is made is interesting. They come from the pot in the shape of a white, luminous lump of melted glass about the size of a hen's egg—perhaps a little larger—on the end of the iron rod. This is rapidly expanded by blowing to a globe the size of a cocoanut. Then the workman rolls it over and over on a cast iron block which lies in front of the workman's seat. This is what gives to the bubble the needed cylindrical form. With a pair of large tongs, something like sugar tongs, the whirling object is altered in shape at the will of the workman until the upper portion of the globe begins to assume the appearance of the upper part of a shade, the glass being kept in a mobile condition by being frequently returned to the furnace and held over the flames. With an instrument for the purpose the material at the extremity of the globe is cut part way through as it whirls upon the end of the rod. It is then held perpendicularly, and a blow jars off the dissevered end, leaving a hole in the globe two or three inches in diameter. Again the globe is heated and then the tongs are used to gradually widen the aperture until the globe is no longer a globe but a shade. This is done so quickly that the spectator does not realize that the work is done before the rod is again held perpendicularly and another jar to the rod causes the shade to drop off and on to what looks like a snow shovel which a small boy holds just under it to catch it. Then it is carried off to the annealing oven. The whole operation has taken only a few minutes. Other articles are made in about the same way, but for some a mold is used.

> *After the articles are made they sometimes remain in the annealing ovens for a week or more, on sliding shelves which are moved at intervals further and further away from the heat of the furnace from which they start.*
>
> *The cutting is done by wheels, four in number. The ware is first applied to an iron wheel. This removes most of the portion to be taken off. Then a wheel of fine sandstone is called into requisition. This does all the fine work. The polishing is done by a wheel of cork and another of brushes. None of the more fancy work, such as etching by acid and cutting by the sandblast, is done here* [in this department].
>
> *Some of the cutting by wheel is very fine, a single piece of ware not larger than a dinner plate, exhibited in the office of the company has fifteen dollars worth of cutting upon it, and in the New York office are to be seen many similar samples.*
>
> *The articles of colored glass made here are plated. When the article to be colored is of the desired form it is thrust into the melted colored glass and a thin coating forms about it; when cool the two substances have become one.*

Six weeks later came a recant of sorts, on June 28, 1879:

> *For some time past it has been rumored that the glass works would shut down permanently when the usual shutting down time for repairs came in July, on the ground that it has not been a paying establishment from the beginning; but it is now positively stated, by one who ought to know, that the company will close for four weeks, commencing next Saturday, for necessary repairs. In one department hands will work until two weeks from Saturday, so urgent are some of the orders. From what can be learned at present the Meriden Flint glass works are in a prosperous condition.*

Trade journals such as *American Pottery and Glassware Reporter* provided informative descriptions of some Meriden Flint Glass pieces, such as this one on July 10, 1879: "Handsome sconces at the Meriden Company's, in New York, are in black glass with relief of white female figures. The candlesticks on each side are ornamented with the pictures of cards. These sconces are so arranged that they may be hung or used as a mantel or table ornament."

The Meriden Britannia Company's head designer at this time was a man named Thomas Pairpoint, considered to be one of the country's foremost silver designers. In 1879, the Mount Washington Glass Company set up a metalware factory to make plated Britannia products to complement

Meriden Britannia Company candle sconce with Meriden Flint Glass white-enameled female figure on black glass. *Dr. and Mrs. Sherwin Borsuk Collection.*

its glassware. It induced Mr. Pairpoint to come as factory superintendent. In August 1879, Thomas Pairpoint resigned from the Meriden Britannia Company, along with about fifty other workers, and the Pairpoint Manufacturing Company was born, modeled after the successful Meriden Flint Glass Company/Meriden Britannia Company partnership.

There was no mention of this exodus of workers from the Britannia works in either the newspaper or Mr. Bourne's journals. The *Daily Republican* noted on August 7, 1879:

> *Work will be resumed in all departments of the glass shop next Monday. Superintendent Bourne has just returned from New York state, where he has been looking up improved machines and models. A larger complement of hands than ever will be employed, and the work promises to surpass in quantity and quality anything that has heretofore ever been put upon the market.*
>
> *Book-keeper W.O. Butler has been confined to the house for the past four or five days with the prevalent malaria.*
>
> *The wives of the glass cutters, who had the picnic at Camp Cedars, last week, visited the same place yesterday, and spent a very pleasant day.*
>
> *Many of the residences in the vicinity of the glass works are being greatly improved by newly laid walks and handsome shrubbery. This part of the city is evidently destined to become one of its most attractive parts.*

The Meriden Flint Glass Company also exhibited its wares that year at Australia's first World Fair, the Sydney International Exhibition. The exhibition opened on September 17, 1879, and ran for six months. There was no mention in local papers about the glassworks' entry, although the company did win a medal for First Degree of Merit, Highly Commended.

On September 17, 1879, the *Daily Republican* noted the following:

> *The clam bake at the organ works last night was attended by about 175 persons. The supper was under the skillful management of Mr. F.N. Hastings with his corps of efficient waiters, and did credit to his ability as a caterer. An original poem by Mr. G.W. Woodruff was read by that gentleman, and abounded in witty allusions to the various employees, whose names afforded opportunities for some very graceful puns. Mr. John Hastings, employed at the Meriden Flint Glass works, favored the gathering with an original poem upon "The Clam," very finely rendered. Supper cleared away, dancing and social chat filled up the hours until past eleven o'clock, when the lights were extinguished, quiet reigned and the second annual clam bake of the Wilcox & White Organ company's employees was a thing of the past.*

A few weeks later, in October, the same paper reported on an incident:

> *A boy who works at the Glass works for 70 cents a day and has no parents has been in the habit of sleeping in a barn and taking his meals with a private family. The other night as he reached the loft of the barn some*

one seized him and threw him down to the floor and his shoulder was dislocated by the fall. It was supposed to have been a tramp. The boy went to the family with whom he boarded and they took him in and sent for Dr. Newport, who came and put the shoulder in place.

In response to the establishment of the American Flint Glass Workers Union, glass manufacturers found it necessary to form their own organization. Just as Meriden Flint Glass Company employees had participated in the establishment of the workers' union a year earlier, the administrators of the Meriden glassworks were involved in the formation of the Eastern Flint Glass Manufacturers Association. These two groups would later initiate the establishment of collective bargaining in the United States.

The *Daily Republican* of November 7, 1879, reported the following:

The association of Eastern Flint Glass Manufacturers held its preliminary meeting at the Astor House in New York city, on Nov. 5th, at which all the prominent Eastern glass works were represented. A regular organization was formed, which is to hold monthly meetings hereafter. The officers elected were Mr. James Gillinder, of Gillinder & Sons, Philadelphia, president, and Mr. George E. Hatch of the Meriden Flint Glass works, secretary. An executive committee was appointed, consisting of W.L. Libby of Boston, Mr. Hibbler, of Hibbler and Dorflinger, Brooklyn, N.Y., F.S. Shirley of New Bedford, Mass., James Gillinder of Philadelphia, and George E. Hatch of Meriden. Steps were taken during the meeting to advance the price of all opal goods about 25 percent.

On the last day of 1879, Thomas Alva Edison first publicly demonstrated his electric incandescent light bulb in Menlo Park, New Jersey.

CHAPTER 5

1880

The new year started out on a less than positive note.

> *Friday, Jan. 2nd, 1880*
> *T. Collins* [Barney McCord's servitor] *struck a boy—Parker—and was discharged.*

On February 6, 1880, the *Daily Republican* noted that "in the police court, this morning, [glassworker] James Callahan was up for making an assault upon Mr. Stretch, the night gate tender at the Main street crossing, and his case was continued until to-morrow."

> *Friday, February 20, 1880*
> *Pay Day—$75.*
> *J. Thompson's Troupe up at the works to day.*
> *Sarah Wilkinson broke valuable ware &c.*
>
> *Thursday, February 26th, 1880*
> *In the afternoon, went with Mrs. Hastings to see her Husband, at the insane hospital, Hartford.*
> *City teams took away 20 load of ashes to mend the side walk on Griswold Street.*
>
> *Tuesday, March 2*
> *Put up the following notice—"Men leaving the employ of this company are*

> *hereby notified that they must give two weeks notice to that Effect."—Per order of the Co.*
>
> *Wensday, March 3rd, 1880*
> *Beautifull day. Very wet night. Came home in the rain, from meeting* [at the Progressive Lyceum].

The *Daily Republican* of March 4, 1880, reported:

> *Owing to the heavy rain storm, Wednesday evening, the attendance at the Progressive Lyceum was not as large as it doubtless would otherwise have been. Yet there was quite a number present, and an interesting discussion was spiritedly conducted, and in the best natured way... Mr. Bourne's essay on "How to best observe the Sabbath" was a very fine paper, and was in accord with the spirit of observing it as a day that should be set apart for the greater glory of God and the benefit of man—a man using his conscience to a reasonable degree to determine how that could be done, in accordance to his taste and with public propriety. The meeting closed by appointing Messrs. Joseph Bourne, E.B. Everitt and Edwin Dayton a committee to devise means of engaging the hall permanently, and perfecting an organization. The next meeting will be held Wednesday evening next, at 7:30 o'clock, in the same hall.*

Bourne continued in his journal:

> *Friday, March 5th, 1880*
> *At noon came home sick, did not return again to glasshouse.*
>
> *Monday, March 8th, 1880*
> *Sick.*
> *Received at glass house 3 kegs of Litharge* [red lead].
> *Sent for Dr. Newport.*

On March 9, 1880, the *Daily Republican* noted that it regretted "to hear that Mr. Joseph Bourne has been confined to his house for the past week, and is threatened with pneumonia. Mr. Bourne has been taxing his energies recently, and took cold, while on his way home in a rain storm from the Lyceum meeting. His large circle of friends hope for his speedy recovery."

A few days later, on March 15, the same paper added: "Mr. Joseph Bourne is still confined to his bed, callers being denied the privilege of seeing him, by instruction of the physician. Mr. Bourne, however, is improving and it is thought that the worst features of his illness are over."

A few days later still, on March 23, the paper announced a positive turn: "Mr. Joseph Bourne, we are very glad to note, is on the mending hand, and now sits up, although probably he will not be at business for a few weeks."

Thursday, April 1st, 1880
Sick. Went out for the first time since March 5th. Went down town and over to the glass house.
Very beautiful day.
Mr. Lyman staking out the land for his new house.

Although he was back at work, Mr. Bourne continued sick for many more days, through which he tried to continue working:

Friday, April 2nd, 1880
Went over to glasshouse. Very beautiful day. Sick.
Pay day—$60.00.

Saturday, April 3rd, 1880
Sick.
Proofs look good.
Poor coal, eye of furnace choked up with iron slag. Got it out, and furnace burnt better.
Very dull day.

Sunday, April 4th, 1880
Sick. Furnace running good. Proofs looked well. Very dull day. No. 1 pot broke, opal.

Monday, April 5th, 1880
Sick. Weather clear. Glass opened plain and good colour. Over at the glasshouse for an hour.

Tuesday, April 6th, 1880
Sick. Fine day. Over at glasshouse felt a chill coming on—came home—very sick in the afternoon.

Wensday, April 7th, 1880
Felt better, over at glasshouse—came home and in the afternoon had another chill. Saw Dr. Newport, he prescribed some powders, took one in the evening—No. 69017—Dr. Newport, from Wilcox & Co.

Thursday, April 8th, 1880
Much better. At home all day. Took 3 powders, felt much better through the day.
Evening set No. 1 & 3 pots.

Friday, April 9th, 1880
Much Better, over at glasshouse, morning and afternoon.
Went to ride with G. Wilcox.
Beautiful day.

Monday, April 12th, 1880
Sarah Wilkinson commenced at noon to clean molds. [See February 24, 1880.]
Mrs. Lutz washing pickles.
$107.20 collected for John Hastings.

George Wilcox was the twenty-four-year-old son of Horace C. Wilcox. He later took over as president of the glassworks upon the retirement of Mr. Lyman.

George's son, Roy, wrote in 1975, "I can well imagine him shunted from pillar to post…to learn what was going on…I've heard him speak in the old days many times of the glass shop…and particularly his friend, Charlie Monroe."

Saturday, May 22nd, 1880
Gave Mr. G. Wilcox 2 pieces of glass to take to Europe with him. He & Bradshaw and Charles Monroe going to Europe.
Cutters off to day, a partial strike.

Monday, May 24th, 1880
Cutters off to day, a partial strike.

Tuesday, May 25th, 1880
Very hot to day.
Mr. Hatch & I went to see Mr. Wilcox about the cutters.

On May 26, 1880, the *Daily Republican* reported that a "large crow blackbird was found this noon in the swamp north of the Glass Works, with a fresh water clam or mussel hanging to its foot, having been caught by the clam while walking in the swamps. It had evidently been confined for sometime, as the clam was dead. The clam and bird are now at the Glass Works."

The building of pots was a tedious and time-consuming operation, taking a number of months to create one pot. They had to be made of a special refractory fireclay capable of resisting the intense and prolonged heat of the glasshouse furnace, and also capable of enduring the sudden contractions and expansions that occurred, as well as the chemical action of the glass matrix they contained. The clay best suited was Goebel's German clay, or clay from Stourbridge, England. After being dried and pounded into a fine dust, the clay had to be exposed to the elements and allowed to "ripen" for about a year. Then a quantity of broken pots was cleaned, ground fine and mixed with the ripened powdered clay to the amount of about one-fifth by weight. Water was added to make a paste, which was "worked" with bare feet to the consistency of putty and allowed to ripen for another six months.

The pots were then carefully built up and shaped by hand, a process that took several weeks. The ripened clay mixture was formed into long ropes, from which the pots were built up layer by layer. All air bubbles had to be expelled from the clay, as these would expand in the heat of the furnace and burst the crucible. The edges had to be covered with wet cloths and allowed to dry between each layer. During the building and afterward, the pots were stored in an airtight room in which the temperature was regulated to about sixty degrees Celsius. After completion, pots were dried in storage again, for four to eight more months. A good stock of finished pots was kept on hand to be used as needed.

Tuesday, June 1st, 1880
Glass all opened good and plain.
John Burke 2nd commenced with M. Collins [pot-maker] *to pound clay.*

Thursday, June 3rd, 1880
Beautiful morning.
Mr. James Gillinder here, very much interested. Mr. Lyman took him around in the afternoon.

Thursday, June 10th, 1880
Making 12 in. trays.
Had some little trouble with men on working &c.

The *Daily Republican* of June 16, 1880, reported on a new construction in town:

> *Of course, everybody in town knows that Mr. and Mrs. Wm. W. Lyman are erecting a new and elegant mansion on the beautiful plot west of the comfortable home now occupied by Mr. Lyman and family, in which they have had so many pleasant re-unions for the past 35 years. The frame of the new residence is now up, and Mr. and Mrs. Lyman thought they would revive the popular custom of "ye olden time," and have a "raising." About one o-clock, a large number of their old-time friends and acquaintances gathered on the lawn that divides the new house and Mr. Lyman's present residence. Carriages came stringing along all the afternoon, every departure being replaced by some new caller, who came to pay their respects, and lend a neighborly hand, in getting the roof in position… They were received by Mrs. Lyman and her daughter, Mrs. Henry Warren, assisted by Mr. Lyman and Mr. Warren…A new and spacious workshop, constructed of new lumber throughout, just back of the house, was turned into a refreshment room, where an excellent lunch was served, consisting of cold cuts, confectionery and lemonade…After partaking of refreshments, the company looked over the new mansion, which is a beautiful and attractive structure, with a solid and substantial foundation. When finished, it will have no superior in appearance and beauty in this vicinity.*

Bourne mentioned it as well:

Wensday, June 16th, 1880
A party at Mr. Lyman's to celebrate the raising of house.

The *Daily Republican* reported on July 7, 1880, that the "Glass Works will shut down Saturday night until August 2nd, and during that time quite extensive repairs and improvements will be made."

On July 31, the *New York Times* noted the following:

> *The Secretary of State has just received a communication from Augustus Morris, Honorary Commissioner and acting executive officer to the United States Commission at the Sydney International Exhibition, embodying a*

> *complete list of the awards made to United States exhibitors. There were 223 exhibitors from this country, and 288 awards were made to them.*
>
> *Highly Commended.—…Meriden Flint Glass Company;…E. Miller & Company, Meriden;…*
>
> *Commended.—Meriden Britannia Company;*
>
> *(and 220 others…)*

Bourne returned to his diary:

> *Monday, August 16th, 1880*
> *Ino Grady commenced to work.*
> *Notified Haden that his pay from to day would be $1.50 pr. move.*
> *Grady wanted $1.75 pr. move. Would not give it.*
>
> *Tuesday, August 17th, 1880*
> *John Grady not at work, left the place.*
> *McNamee refused to blow mustards and left this afternoon.*
> *William Martin arrived from Philadelphia—tumbler gaffer.*
> *John Nonnenmacher commenced to work this morning, in mixing room.*
> *William Martin commenced this morning.*
> *Timothy Warren commenced as sticker up.*
> *Delaney gave notice this morning, going to Libbey. Sent 4 dollars to Coffin to pay his fare.*

On September 9, 1880, *American Pottery and Glassware Reporter* noted that "Hogg & Patterson, 17 West Houston Street, New York, have now in store a splendid assortment of gas and kerosene globes, street lamps, etc., embracing all the latest and most popular styles, both plain and decorated. They are the agents for the goods made by the Providence Gas Burner Co., and the Meriden Flint Glass Co., so well and favorably known throughout the country."

The *Daily Republican* of September 30, 1880, reported that a "very clear presentation of the Tariff question, and its effect upon Meriden industries, is printed in this issue, from Mr. Joseph Bourne, an intelligent mechanic, who has worked both in England and here, and who belongs to the party known as independent voters, but who now stands with the Republican party, because it discriminates in favor of American labor."

Two weeks later, on October 16, the paper noted that General Grant was in Hartford and that an immense crowd of veterans and boys in blue were in the procession; thousand of visitors occupied the city, including 1,500 from Meriden.

Saturday, Oct. 16th, 1880
Nobody working in glasshouse. General Grant at Hartford.
Pay Day—$50.00

Going home with "the shakes" was not always due to alcoholism or a manifestation of "blood poisoning" from working with chemicals.

Tuesday, Oct. 19th, 1880
McCabe off sick with the shakes, went home sick at noon.
Whallen sick.
Was sick myself in afternoon.
D. Forbes sick.

Wensday, Oct. 27th, 1880
McCabe sick with the Malaria.
M. Hoar went home to get rid of a cold. Gave him 25 Dollars.

Mr. Bourne, an ardent Republican, kept track of his employees' party affiliations:

Tuesday, Nov. 2nd, 1880
Beautiful Day.
Election Day for President, Vice President, members of Congress and state elections throughout the country.
Ino Burke voted Democratic.

Wensday, Nov. 3rd, 1880
Gen. Garfield and the Republican Ticket elected.

Chapter 6

1881

The new year started with a few discharges at the company.

> *Saturday, Jan. 1st, 1881*
> *Very cold beautiful day.*
> *Discharged the following persons, to leave Jan. 8th:*
> *G. Carter, Samuel Haden, Anderson, D. Martin.*
> *Changes made in shops, &c.*
> *Discharged Sarah Wilkinson to day.* [See February 24 and April 12, 1880.]

> *Monday, Jan. 10th, 1881*
> *The hardest morning known for years &c. Discharged 4 cutters in shop*—[James D.] *Bergen, Mahary, Hack, & Haden.*
> *Martin & Anderson went off this forenoon.* [See January 1.]
> *Ino Collins off.*

"Mr. Geo. Hatch started for Philadelphia to-day, to attend the meeting of the national association of glass manufacturers, which begins in that city tomorrow," noted the January 11, 1881 *Daily Republican*.

William Worcester Lyman was an enterprising and public-spirited man. Beside being an inventor (he is credited with inventing the modern twist can opener, among other things) and a director of the Meriden Britannia Company, he was also at different times president and on the boards of the Meriden Cutlery Company, the Meriden Horse Railway Company, Meriden National Bank and City Savings Bank—he was also, of course, president of the Meriden Flint Glass Company. He was a member of the Common Council, Fifth Ward, and he served as Meriden's representative in the Connecticut legislature in 1859 and 1881. On January 12, 1881, the *Daily Republican* reported that the "Hon. W.W. Lyman made his first appearance in the House to-day, and took the oath. He is still weak, although much better than he has been."

Tuesday, Feb. 8th, 1881
Cutters out, have to come in on time.

Wensday, Feb. 9th, 1881
Cutters out until noon.
All the glassmakers off to day.

Thursday, Feb. 10th, 1881
All hands working.
Thaw continuing with fog &c.

On February 15, 1881, the *Daily Republican* reported that "the new residence which is being built for Hon. W.W. Lyman on Britannia street is progressing rapidly, and it is now the intention to have it ready for occupancy by the first of May. Mr. John F. Butler is doing the finishing and painting of the house, and it will be very handsome when completed."

Mr. Lyman's "new" house still stands today, next door to his previous homestead. To the left of that is a house built by Mr. Lyman for E.G. Pepper, an engineer at the glassworks. Across the street from these is the Mansard-roofed apartment house in which Mr. Bourne resided with his wife and two younger daughters, Mae and Emma.

All of these buildings, which were specifically built to house glassworkers, are still standing and occupied today. They are a short walk from the glassworks site.

Wensday, Feb. 23rd, 1881
Beautiful day.
Making ruby, green & blue plated ware.
No fluor spar. Made up a batch XC—XX.
Commenced to draw coke from the Meriden Gas House.

Coke was a byproduct of the coal gasification process. The Meriden Gas Company produced gas by heating coal to a very high temperature, which broke the coal down into three components: gases, tar and coke. Coke was a necessary ingredient for glassmaking, and the Meriden Flint Glass Company purchased some of its coke from the gas plant.

In 1881, it was thought that glass might be a good material with which to make toilet bowls. Glass was regarded to be stronger than porcelain and therefore less likely to break. Further, glass did not become discolored as porcelain did. It was hoped that "the manufacture of annealed glass may be soon improved sufficiently so as to allow of making glass water-closets as well as other glass plumbing fixtures."[10]

The Meriden Flint Glass Company began making glass toilet bowls in March 1881. A patent for the water closet bowl was issued to Mr. Hatch in May.

Friday, March 4th, 1881
Very wet rainy day.
Making Water Closets.
Garfield inaugurated to day.

On March 5, 1881, Joseph Bourne addressed the spiritualists at the GAR Hall, according to the paper.

Friday, March 11th, 1881
Cold blowing day.
McHugh & Chase commenced on 4-inch Smoke bells—and left off.
Mr. Sheer here from Boston & Sandwich.
Lutz went home with the shakes.
Wetherbee run a nail into his foot through the night.

The Westland safety lamp was developed in 1878. It featured a double-walled fount, or fuel chamber. The inner chamber held the lamp oil, which was burned to provide light. The surrounding chamber was filled with carbonic-acid gas and was intended to extinguish a fire if the lamp fount was broken.

On June 7, 1881, George Hatch of the Meriden Flint Glass Company was issued a patent for an improved method of manufacturing the double-walled fount used on the safety lamp. The company had worked on perfecting the fount mold for several weeks before filing the patent application.

Wensday, March 16th, 1881
Beautiful day.
McCabe making 12¼ by 14½ globes.
Tried the [Westland] *fount mold to day.*

Wensday, March 23rd, 1881
Cold raw day.
Pay $3.50 cts. for Bbl. of Cider had 4 years ago to Farmer Bacon, Westfield.

On March 30, 1881, the *Daily Republican* reported that "the carved mantels for Mr. W.W. Lyman's new house have arrived from Bridgeport, and are being put up to-day. They are of wood to correspond with the wood in the different rooms, and are very elaborate and handsome."

Daniel Forbes was in charge of the cutting department, where newly made glassware was polished and cut. Glass was polished with a paste, or putty, composed of tin and lead applied to a rapidly spinning wheel. A considerable quantity of this was thrown off as dust, which was very injurious to the health of the workers.

Mr. Forbes was developing a new kind of polishing putty that would be less harmful to inhale. A patent was granted for his invention on August 1, 1882.

Saturday, April 23rd, 1881
Built Forbes a furnace for putty making.
George Wilcox made up 2 Common and one Opal Batch.

Monday, April 25th, 1881
To day reduced the following parties in their pay—to take effect from May 1st, 1881:
J. Moore from 1.50 to 1.40 per move
O. Hickenbottom 1.50 to 1.25 per move.
J. Car 1.25 to 1.15 per move
W. Hobson 1.35 to 1.25 per move.
J. Ray 1.87 ½ to 1.75 per move
J. Holdbrook 1.25 to 1.15 per move
M. Collins $13.50 per week to $12.00 per week
J. Collins $13.50 per week to $12.00 per week
P. Whallen 15.00 per week to $12.00 per week

Thomas Wilkinson $1.50—$1.25
J. Day $1.25—$1.00
Warren $1.25—$1.00
Conway $1.25—$1.10

On May 2, 1881, the *Daily Republican* noted that "the Meriden Flint Glass Co. are making some beautiful lamp shades for the Profile House in the White Mountains. The decorations embrace fine representations of the Profile House, and the 'Old Man of the Mountains' on Mount Cannon. Both will be readily recognized."

Profile House was a grand hotel located in Franconia Notch, in the White Mountains of New Hampshire. It could accommodate six hundred guests and flourished as a summer resort before burning down in 1923. Today, the parking lot for the Cannon Aerial Tramway is located at the site of the Profile House.

The "Old Man of the Mountain" was a series of granite cliffs that, when viewed from the right angle, looked like a jagged profile of a man's face. Also known as the Profile, it collapsed on May 3, 2003.

Monday, May 2nd, 1881
Glasscutters not working to day. Glassmakers committee waited on me, to protest against the reduction of the pay of some of the men.

On that day, twenty-five Meriden Flint Glass Company glass cutters went on strike because of the wage reduction. Labor relations at the company had been strained for some months at this point.

The plant did not close down during this twenty-eight-day walkout. All of the striking glass cutters were replaced with new men, at the same rate of pay as before the strike.[11]

To some extent, it was "business as usual" as the factory continued to produce glass. Although some of the strikers engaged in malicious activity, Mr. Bourne seemed to gloss over it, perhaps trying to put a positive spin on a bad situation:

Tuesday, May 3rd, 1881
Beautiful day.
Cutters out.
Set No. 8 & 4 Pots.

There were many children employed at the Meriden Flint Glass Company. Most worked as assistants to the men, and some were glassmaking apprentices. The typical length of an apprenticeship was five years, starting around age fourteen.

In 1880, a fourteen-year-old boy named Philip, unbeknownst to his mother, began working at the Meriden Flint Glass Company as a glass decorator. For the first six months, he received no pay whatsoever. Five years later, at the end of his apprenticeship and with the glassworks foundering, Philip Handel opened his own glass decorating firm with partner and glass company co-worker A.G. Eydam. After buying out Eydam's share of the business, the Handel Company continued to operate for many years, turning out beautifully decorated glass and metalware that are eagerly sought by collectors today.

Connecticut labor laws at this time stated that no child under the age of fourteen was to be employed in any business unless she or he had attended some type of day school for at least sixty days during the preceding year, for at least six consecutive weeks. Employers were required to keep a certificate of school attendance for each child under age fourteen employed. Children as young as twelve were known to have worked at the glassworks.

Families who failed to send their children to school for the required time were fined five dollars for each week missed, a substantial amount of money in 1881, especially for a destitute family who found it necessary to send the children out to work.

No child was allowed to work more than ten hours per day or fifty-eight hours per week, under a penalty of fifty dollars. School committee members visited the factories at least once per year to ensure compliance with the law.[12]

Wensday, May 4th, 1881
Committee man here enquiring about the ages of the boys and their time of going to school.
Beautiful day.
Cutters out.

Monday, May 9th, 1881
Cutters still out.

Labor problems were plaguing the entire glassmaking industry at this time. It was common practice for glass manufacturers to communicate with one another regarding terminated employees who might be seeking jobs elsewhere, effectively blacklisting them. In 1880, Thomas G. Hawkes opened a glass cutting shop in Corning, New York. The Corning Glass Company supplied its glass blanks for cutting. The Meriden Flint Glass Company's secretary and sales agent, George Hatch, wrote the following letter to Hawkes regarding Meriden's striking glass cutters on May 11, 1881. Mr. Hatch's use of the phrase "Society men" refers to the fact that the strikers belonged to the American Flint Glass Workers Union.

> *I enclose a list of 15 Glass Cutters, all of whom are Society men, and have given us constant trouble. We changed our system of work, and contracted with the old foreman to do our cutting and these Society men refused to work for him. They were, of course, discharged by this Company, but have been exceedingly malicious and done what they could to prevent our getting men. We shall consider it as a mark of courtesy if you see fit to refuse to employ them.*[13]

On May 11, 1881, the *Daily Republican* noted that "Mr. Hatch reports from the Glass Factory that they will have six first class cutters who have gone to work to day and he had not trouble in getting all the men the company required. They have twelve cutters in all now at work."

Two days later, the paper reported that "William Wilkinson, one of the employees at the Glass Works, accidently stepped in a tub of hot water, Thursday afternoon, and burned his leg so badly that the skin peeled off. Dr. Mansfield attended to his injuries." On May 16, the paper noted that "the United Flint Glass Cutters Mutual Assistance and Protective society will have a picnic, at Terrace Garden, on Saturday of this week." The picnic site was located on Columbus Avenue in Meriden and is still there in 2011, although much diminished.

> *Wensday, May 18th, 1881*
> *Very warm.*
> *New cutters came.*

Twenty-five cutters in all were replaced as a result of the strike.

> *Thursday, May 19th, 1881*
> *Taken very sick forenoon.*

Friday, May 20th, 1881
Sick—Malaria, Dysentery—had Dr. Newport.

The *Daily Republican* of May 28 reported that "Mr. Joseph Bourne is again ill, he having such another attack as he experienced a year ago. He has been confined to the house for the past few weeks. It is pleasant to know, however, that he is recovering, and able to sit up, although not able to be out of doors."

One day after the cutters' strike ended, Mr. Bourne returned to work:

Tuesday, May 31st, 1881
Fine day over at Glasshouse.

The Connecticut Bureau of Labor Statistics published a report in 1886 that compared, among other things, the effects of employee strikes and employer lockouts in the state in previous years. The following information is recorded about the May 1881 cutters' strike at the Meriden Flint Glass Company and provides some interesting details about wages and working conditions.

Strike—1881
Cutters, Meriden Flint Glass Works, for "increase of wages."
Number of flint glass manufacturers in the country at that time—122
Ordered by Labor Organization—No
Number of Establishments—1
Days closed—none
Beginning—May 2, 1881
Ending—May 30, 1881
Duration—28 days
Succeeded—no
Number of employees striking—25 cutters
New employees after strike—25 cutters
Brought from other places—25
Weekly working hours before strike—60
Weekly working hours after strike—60
Daily cutter's pay before strike—$2.75
Daily cutter's pay after strike—$2.75
Employees' Loss—$1,626.00
Employees' Assistance—none
Employer's Loss—$375.00

Number of Male employees before the 1881 strike—150;
Number of Male employees after the 1881 strike—150;
Number of Female employees before the 1881 strike—4;
Number of Female employees after the 1881 strike—4;
Total number of employees before the 1881 strike—154
Total number of employees after the 1881 strike—154
Average Daily Wage Male/Female before the 1881 strike—$1.75/.80;
Average Daily Wage Male/Female after the 1881 strike—$1.75/.80.[14]

Bourne returned to his diary:

Wensday, June 1st, 1881
Very warm day.
Meeting on the Water Closet Affair in Meriden B. Co. office.

The water closet affair was detailed by the *Hartford Daily Courant* on June 3, 1881:

The recent advertisement for investors in an enterprise for the establishment of a glass furnace in this city [Hartford] *was not responded to very freely, and the parties interested in the movement have made other arrangements…The proposition was to manufacture the opaque glass used in the "Hartford Glass Water Closet" recently patented by Mr. Matthew Hogan, plumber, of this city…However, the project has been abandoned, and the company will for the present procure its supply of glass from the Meriden Flint Glass company, with which a contract was made this week for 40,000 pieces.*

Bourne continued his diary:

Monday, June 6th, 1881
Glass plain.
Commenced to make the Water Closets—to make at least 120 Bowls.
Walker off, hands swelled up.
Farrer discharged, for stealing Coal.

On June 27, 1881, the *Daily Republican* presented an article on a bit of finished construction:

MR. LYMAN'S NEW HOME
DESCRIPTION OF THE INTERIOR AND THE GROUNDS

The mansion just completed by Hon. Wm. W. Lyman, on Britannia street, is undoubtedly among the most elegant and best arranged of the attractive residences in this vicinity.

On entering the massive front doors, a broad and unobstructed hall indicates that everything is spacious. To the left as you enter is the parlor, which is finished in curled maple, with bird's-eye panels, highly polished, and without a particle of it painted. The ceilings are elegantly frescoed in delicate tints, blended here and there with strong colors, producing a striking effect. The appointments in this room are rich in every particular. The carpet is moquet [a kind of carpet with a thick, nappy surface], *harmonizing with the frescoing. The furniture is old gold pattern of brocaded satin damask. Oil paintings of Mr. Lyman's mother and Mrs. Lyman's father, and a fine landscape scene of St. George, by Moore; a Hudson scene, and a view of Kensington by the same artist, are the most prominent wall decorations, and a few panels and ottoman covers, fancy work in the kensington stitch, executed by Mrs. Warren, add to the beauty of the parlor. The cornices and poles are heavy gilt, and the large mirror of plate glass standing between the two main windows, reaching from the ceiling to the floor, are richly ornamented with gilt crown, and a base heavily carved.*

On the right, and directly across the hall from the parlor, is the sitting room, and while the designs differ from the parlor in the frescoing, they are equally attractive, and this room is furnished as elaborately, the modern custom being observed of furnishing no two rooms alike. In both the parlor and sitting rooms are massive cabinet mantels of polished ash and maple respectively, carved in delicate pillars as they ascend, and tapering down on either side to the main shelf, where they are finished with projecting mirrors of beveled glass, on hinges, that serve for jewel caskets or kindred purposes. The hearths in these rooms are in keeping with the rest of the appointments, while in the sitting room hearth is a pair of brass andirons, 100 years old, as finely polished and free from blemish as if they came to-day from the foundry.

West of and connecting with the sitting room is Mr. Lyman's library, a cosy nook, and from it one gets a full view of both sitting room and parlor across the broad hall, and the crystal gasaliers that are in as straight a line in sitting room, hallway and parlor, as if they were soldiers on dress parade, give a fine impression of the richness and elegance of the appointments.

The doors of these rooms, it should have been said, are heavy and polished to a gloss, and stained glass of dead crimson, relieved by decorations of tricolors, ornament each door as well as the main entrance.

Off the library, separated by a passage, is Mr. Lyman's bed room, and it is fitted up with a view to comfort and convenience.

Across the hall is the dining room, which is finished in polished oak, is perhaps the finest piece of artistic work in the house, and we believe nothing is hazarded in saying that the dining room of no private gentleman in the state will go ahead of it... The ceiling is frescoed with duck, lobster, game, fish, grape vines, &c, the greatest care being taken to introduce characters that gave the artist ample room for blending and harmonizing colors. A rich gilt molding on the walls relieves the shades of this frescoing and bring them into agreeable prominence, without any attempt to force them. The imported mantel, of polished tile, is surrounded by a grape vine similar to the frescoing, rigidly observing harmony... The paintings in the dining room are the work of a local artist, Mr. F. Miller [of the glassworks], *and he faithfully represents fruit, fish, game, fowl, &c. One piece is particularly noteworthy, being a dish of strawberries, transferred to canvas from Mr. Lyman's strawberry beds.*

The butler's pantry adjoins the dining room, and is also of oak, and fitted up with every known convenience. The kitchen is divided only by a door from the pantry, and comfort and convenience have been studied in the arrangement of it.

The Second Floor is reached by an elegant staircase, in full view from the main entrance, but erected in the extreme rear, so as not in the least to obstruct the spacious hall or the passage ways. This staircase is of natural ash. It is massive in construction, and the angle at which it is built makes it easy of ascent.

A recess to the left above the broad staircase is utilized to good advantage for a stained glass window, the decorations and colors of which happily harmonize with the frescoing above and about it, and the entrance below, and breaking the luxuriant monotony of the continuous frescoing, without being at all abrupt to the eye. The frescoing is carried to the ceiling of the second floor, and continued through the hallways there.

On this floor is Mr. and Mrs. Warren's suites of rooms, finished with polished panels of wood work, the decorations and frescoing equally agreeable as the rooms already described.

Miss Etta Warren's room is also on this floor, adjoining her parents, and is fitted up elegantly with a light blue carpet, and other appointments to match. Miss Etta has an oil painting of her favorite kitten among the paintings, which was executed by F. Miller.

The great chambers, of which there are two, are also on this floor, and, of course, are equally well finished as the other rooms. They are finished with crimson suits.

In the center of the second floor, having a view of the mountain sweep, is a recess, which is Mrs. Warren's sewing room, perhaps the most romantic spot in the whole house, for its charming view of the surrounding scenery, as it commands a view of "hanging hills."

On the upper floor are the billiard room, smoking room, music room, Miss Etta Warren's play room, and a number of rooms used for storage, &c. On the garret also is a large tank where surplus water is stored for emergencies.

Mr. Lyman's "new" house, circa 1890. Mr. and Mrs. Lyman are seated on the porch. *MHS.*

Mr. Lyman's carriage house, 2011.

On July 2, newly elected president James A. Garfield was shot and mortally wounded at a railroad station in Washington, D.C. He was shot twice, in the back and arm, by a lawyer, Charles J. Guiteau, who had unsuccessfully sought an appointment to the U.S. Consulate in Paris.

On July 14, outlaw William H. Bonney Jr., alias "Billy the Kid," was shot and killed by Sheriff Pat Garrett in Fort Sumner, New Mexico.

The *Daily Republican* of July 16, 1881, announced that "the Glass shop will shut down, Saturday, for inventory and repairs. The suspension ought to be for three weeks, but business is so brisk that it is thought a week's shut down must be made to answer…Mr. Joseph Bourne, we regret to learn, is again seriously ill, the result of a relapse, caused by going to business too soon. He intends, when strong enough, to take an extended vacation, if he can arrange his duties so as to permit it."

The July 25 paper had an update: "Mr. Joseph Bourne is at the seaside, outside Boston, endeavoring to recover his health, which has been shattered."

The August 1 paper noted that "the Glass shop started up this morning with a full complement of hands, after a two-weeks stop for repairs."

> *Saturday, August 13th, 1881*
> *Returned home on receipt of telegraph Aug. 12, 1881. Furnace had run down and heat could not be obtained to melt the glass.*
>
> *Sunday, August 14th, 1881—Furnace as hot as can be obtained with coal on hand.*

The August 15, 1881 *Daily Republican* reported that "the furnace at the glass shop went out…on account of poor coal, and the hands had a holiday, but will be at work to morrow again…Mr. Joseph Bourne returned to this city Friday evening, his presence being required at the glass shop. His health has not improved at all, and he plans to return to Boston again."

> *Monday, August 15th, 1881*
> *Sent down to Meriden B. Co. for coal—had 2 loads and furnace commenced to recover.*
>
> *Tuesday, August 16th, 1881*
> *Furnace still improving on new coal. 4 loads more received.*
>
> *Wensday, August 17th, 1881*
> *Furnace almost up to regular heat—all hands working on opal—making shades & water closets.*
>
> *Thursday, August 18th, 1881—furnace all right.*
>
> *Friday, August 19th, 1881*
> *Furnace hot and everything in order.*
> *Returned to Boston on the 1:50 train.*

The August 29, 1881 *Daily Republican* noted that "Mr. Joseph Bourne returned to the city Saturday, his health being slightly improved. He was not able to stand the heat of Boston, and he returned home. His friends are pleased to note that even a slight change is noticeable."

Friday, September 2nd, 1881
William Urwin & son left to day to go west.
Glassmakers had a meeting on wages.

Tuesday, September 6th, 1881
Very hot day.
President Garfield removed from Washington to Long Branch.
Very few men working in glasshouse.
Felt very unwell to day.

Tuesday, September 13th, 1881
Beautiful morning, everything going on well in glasshouse.
Prayers offered for the President's recovery to day in all the churches.

Monday, September 19th, 1881
President Garfield died this evening at 10:35 in the Francklyn Cottage at Long Branch.

Wensday, September 21st, 1881
Governor's Day at the Fair—only McCabe's, McHugh's & Henriott's shop working forenoon.
From to day I remained home very sick and did not come to glasshouse untill Friday the 11th day of November.

On September 24, 1881, the *Daily Republican* reported that "in the north part of the city, on Britannia, Griswold and North Colony streets, nearly all the houses show some symbol of mourning. Mr. Lyman's new home in particular is fully draped."

A few days later, on September 27, the paper noted that "Mr. Joseph Bourne is again very ill, his lungs now being troubled, and complicating his other troubles. Mr. Bourne's continued illness will be a source of regret to all who know him, for he is always full of practical sympathy for the afflicted himself."

On October 28, the *Daily Republican* reported that "Joseph Bourne, who has had a long siege of illness, from which it was not expected he would recover, is now able to sit up. He is still weak, but his improved condition has given his friends fresh courage."

The *Daily Republican* of November 3 had glad tidings:

A Birthday Reception

The large, fine new residence of Hon. W.W. Lyman of Britannia street looked very attractive to passers by early Wednesday evening, with a bright light shining through every window upon the mist and drizzle without, and the merry peals of children's laughter that followed each other in quick succession were heard with smiles far up the street. Within the house as gay a scene as could be well imagined brought pleased smiles to the faces of all the older lookers on.

The occasion was the 9th birthday of Miss Etta L., daughter of Henry Warren and grand-daughter of Mr. Lyman, and her charming little reception to her child friends of about her own age was the first formal "house warming" the new residence has received. Certainly no home was ever dedicated with a more joyful gathering. Over a score of bright eyed children romped and shouted through the spacious rooms, from the elegantly furnished parlors to the cosy play rooms under the roof, with no restraint put upon their noisy merriment.

The reception began at 3 in the afternoon, the pretty little hostess meeting all her guests with a graceful welcome, and the afternoon and early evening were crowded with appropriate festivities, music, dancing and games, keeping mind and body active and the jollity sustained. Miss Etta received numerous beautiful birthday presents, including some exquisite little floral gifts, prominent among them being a dainty cradle from New Haven with a body of carnations and garniture of tea and tube roses and pansies. The souvenirs of the birthday party will be many and pleasant.

The November 10 *Daily Republican* reported that "Joseph Bourne, his friends will be glad to know, has so far recovered from his long and severe illness as to be able to turn his attention to business." The next day, "Tremont street is being filled in opposite the Glass works and the road graded up with the cinders and 'clinkers' from the furnaces, which appears to be a very good material for such a purpose."

On November 17, the paper reported on a fundraising collection:

Michigan Sufferers.
The Dollar Fund Subscriptions.

Already acknowledged	*$397.01*
Employees Meriden Flint Glass Co.	*27.25*
	$824.76

Top: George Hatch's inkwell in flint (crystal) glass. *Author's collection.*

Middle: A Meriden flint glass "Crown" shade. The same etched scene is seen on an etched shade in the Mitchell, Vance catalogue. *Author's collection.*

Bottom: Opal fruit bowl with blue shaded exterior, raised gold decoration and applied amber ruffled rim. *Author's collection.*

Above: 1879 Sydney International Exhibition medal. *MHS.*

Below: Pressed optic pattern blue glass nut bowl (left) with colored Coralene decorations, on a silver-plated Wilcox Silver Plate Company stand. Opal glass biscuit jar (right) with shaded pink exterior and raised gold decorations. The silver-plated lid and base of the biscuit jar bear the Meriden Britannia Company's trademark. *Lionel DeRagon Collection.*

Top: Close-up of raised gold peacock on opal fruit bowl, with shaded pink exterior and applied amber ruffled rim. *Author's collection.*

Above: A Degenhart paperweight made for Peter Bottner, a glassblower at the Meriden glass factory. *Author's collection.*

Right: Pressed flint glass vase in Meriden Britannia holder. *Author's collection.*

Left: Pressed flint pickle jar in Meriden Britannia holder. *Author's collection.*

Below: An unfinished round "blank." *MHS.*

Above: Sky-blue jewel box with white enamel decoration, made for the Meriden Silver Plate Company. *Author's collection.*

Right: Two-ring cylinder lamp with an orange, diamond-shaped Meriden Flint Glass Company label inside the cylinder. *Author's collection.*

Five-sided cut flint bowl once belonging to Mollie Callahan Nolan. All surfaces and cuts are hand-polished. *MHS.*

A pressed flint fruit or ice cream dish in the "Tree of Life" pattern. This item is shown in the 1882 Meriden Britannia catalogue. *Author's collection.*

Above: Minton tile trivet with a Meriden Flint Glass Company label on the back. *Author's collection.*

Below: Two perfume bottles and a powder jar. These are shown in the 1878 Meriden Britannia catalogue in their silver-plated stand. *Author's collection.*

Blue craquelle glass vase in a Meriden Britannia Company holder. *Author's collection.*

Holmes, Booth & Haydens lamp with labeled MFG cylinder. HB&H was a Waterbury, Connecticut manufacturer of lamps and burners. *Author's collection.*

> *The above collection from the Glass Shop speaks well for the generosity of the hands employed there. There is no shop in town upon which there are more constant appeals for aid, the subscription papers there averaging three a week for some unfortunate shopmate or other worthy object of assistance. Knowing this we consider their donation to the sufferers liberal, and we thank them for it, and hope the other shops will do likewise. We would like to see every shop in town do something.*

Bourne continued to maintain his diary:

> *Saturday, December 3rd, 1881*
> *Cold rainy day.*
> *Set No. 10 pot at noon.*
> *Felt sick this afternoon. Called in Dr. Mansfield.*
>
> *Sunday, December 4th, 1881*
> *Sick—a return of shortness of breathing.*

On December 14, "Henry Fields, an employee at the Glass works, had a pair of toes on his right foot smashed by the falling of a 300 pound weight last evening," according to the *Daily Republican.*

> *Thursday, December 22nd, 1881*
> *Finished the large punch bowl—cut rich and was a beautiful piece of work in colour and finish.*
> *Received 14 pots from Pittsburgh.*
>
> *Saturday, December 24th, 1881*
> *Shipped the large cut Bowl to Steel Bros., Philadelphia.* [See March 9, 1882.]

The December 29 *Daily Republican* reported that "Michael Collins, employed at the Glass shop, fell on broken pieces of glass, Wednesday, cutting his right hand severely. Dr. Childs was called, and picked out the pieces of glass which stopped the bleeding."

> *Saturday, December 31st, 1881*
> *Clear cold day.*

Shops working in the forenoon.
Tinning the roof of shed.
Taking stock in the afternoon.

This represents the end of 1881, as well as the end of Mr. Bourne's journals.

Chapter 7

1882

The year 1882 began auspiciously enough, with the newspaper reporting the flint glass company running a full complement of hands, full time, every day. There had been some changes in management, with which there seemed to be "very general satisfaction now among the employes at the Glass shop."

The February 3, 1882 *Daily Republican* reported that "a surprise party was given Mr. and Mrs. Henry Fillebrown [MFG glass cutter] on North Colony street Wednesday evening by about fifteen of their friends. The evening was spent in playing games and singing. William Harvey and James Gay sang a duet in a pleasing manner."

The February 2 *American Pottery and Glassware Reporter* noted the following:

> *Mr. Hatch, secretary of the Meriden Flint Glass Co., was in New York last week and reports that his company will bring out several new and beautiful specialties during the coming season. The company* [has] *just completed their arrangements to control the process of photographing on opaque glass, or what is generally known in connection with porcelain decoration as the "Helio-Ceramic Process." The designs are photographed on the glass and will stand being burned in without marring their beauty. The company intend utilizing this great invention in beautifying lamps and gas shades, and will shortly bring out a splendid line of their goods embellished with portraits of eminent authors, actresses, statesmen, &c., and also a full line of mantel vases similarly ornamented. These goods are especially adapted for use in the library, dining room, saloons, &c. If parties choose they can*

have their shades or vases decorated with family portraits. When lighted up the effect of this style of decoration is superb.

Drug addiction was little understood in the nineteenth century. There were no illegal drugs until San Francisco passed ordinances against opium smoking in 1875. Morphine was commonly used to treat alcoholism.

The February 11 *Daily Republican* reported the following:

Many of the hands of the Glass Shop must have known Mr. Henry Whitney, general manager of the New England Glass works, whose untimely death by accidental poisoning is reported by telegraph to-day.

Henry Whitney, who for many years had been connected with the old New England Glass Manufacturing Company of Cambridge, first as clerk and then as general manager, died suddenly Friday in Cambridgeport, his death being the result of accident. He had been absent from his home for six weeks past, under treatment for a disease with which he had long been afflicted, and returned Thursday. He had been in the habit of taking almost daily small quantities of morphine and Thursday swallowed two grains from a paper which he supposed contained that drug. It proved, however, to be strychnine which had been purchased some time ago for the purpose of killing rats. As soon as the terrible fact was discovered, medical assistance was obtained, and five physicians did all in their power to counteract the effect of the poison without avail, death resulting in a short time.

On March 9, *American Pottery and Glassware Reporter* printed the following:

"CORRESPONDENCE"…I have returned here from a trip to Boston and Meriden…I should judge that it is about a mile and a half from this factory [Bradley & Hubbard Manufacturing Company] *to the works of Meriden Flint Glass Co., which I next visited. These works were started some years since in opposition to the prophecies of many "knowing ones," but have turned out in every way a most decided success…The table ware shown me was really elegant in variety and finish, the cut and engraved work being equal to any of the imported.*

Last season they furnished a magnificent bowl of this class, superbly cut, decorated with bunches of grapes, &c., which was held at $250. [See December 22 and December 24, 1881.] *This would hold punch or egg nogg enough for all of a family's New Year callers, and leave some to spare.*

On March 29, the *Daily Republican* noted that "Mr. Joseph Bourne, his many friends will be sorry to hear, does not appear to gain. He is still very weak, but it is hoped that he may build up with the balmy weather." On April 12, the paper reported that "Joseph Bourne's now able to sit up and is rapidly recovering and in a few days will be able to go out of doors." On April 17, the paper noted that "Mr. Joseph Bourne shows but slight, if any, indications of improvement."

On April 21, 1882, however, the *Meriden Daily Republican* had tragic news:

> *The many friends of Joseph Bourne will be pained to hear of his death, which occurred* [Thursday] *night at about half past ten. He had not been in robust health for some years, and his last illness lasted about two months, although he was only for a few days considered dangerously ill. The immediate cause of his death was heart disease. He had been sitting up in bed, and had asked his wife to lay him down, but died while she was complying with his last request.*
>
> *Mr. Bourne had been superintendent of the Flint Glass Works ever since its formation, and was greatly beloved and respected by all its employees and officers as well as by his numerous friends.*
>
> *Mr. Bourne was a native of England and was 56 years of age. He leaves a wife* [Fannie] *and three daughters, Mrs. C.H. Lake* [Henrietta] *of Boston, who formerly resided here, and Misses May and Emma. He was a member of the Knights of Honor and the Sons of Temperance. The funeral arrangements have not yet been completed, but the remains will be taken to Cambridge, Mass., for interment.*

Mr. Bourne was buried at Mount Auburn Cemetery in Cambridge, Massachusetts. Thus ended the Bourne era at the glassworks.

On January 18, 1778, English navigator Captain James Cook reached the Hawaiian Islands, which he dubbed the "Sandwich Islands," after one of his sponsors, John Montagu, Fourth Earl of Sandwich.

In 1879, the islands were still known by that name when King David Kalakua began building Iolani Palace in Honolulu. Completed in 1882, the palace served as the official residence of King Kalakua and his sister, Queen Liliu'okalani, Hawaii's last two monarchs, until 1893.

Iolani Palace is the only royal palace in the United States and is still used today (in 2011) for very formal state functions. Listed on the National Register of Historic Places, it was opened to the public in 1978.

Joseph Bourne's home on Britannia Street, located directly across from Mr. Lyman's new house, in 2011.

On May 1, the *Daily Republican* reported that "the Meriden Flint Glass Company are shipping this week a complete set of richly etched shades and globes to the Sandwich Islands, for the palace of King Kalakua. No expense has been spared on the artistic work, all the various designs of scenery familiar to the cannibals being etched and decorated in the finest manner known to the most skilled artists and mechanics."

The *Daily Republican* of May 13 reported the following:

> *The Meriden Britannia company have just added to their fine stock of goods, which makes the name of the company world-wide famous, some new designs which for elegance and beauty surpass anything before attempted in this country...Still another design of different style is an elegant bowl resting upon a pedestal of polished silver, the arms holding the choice cut glass bowl being formed of rose vines which cluster about it and show wild roses in bloom and in bud. The cut glass bowl was also manufactured in Meriden, by the Meriden Flint Glass Company, and for beauty and clearness cannot be surpassed in the world.*

Above: The "elegant bowl resting upon a pedestal of polished silver" cut by the Meriden Flint Glass Company and shown here in the 1882 Meriden Britannia catalogue. *MHS.*

Left: A chandelier made by the Meriden Flint Glass Co. for Iolani Palace being removed for safekeeping after it was shaken by bombs during the attack on Pearl Harbor. *From the* Honolulu Star-Advertiser.

The sudden death of Mrs. Stephen McCabe occurred less than three weeks after the death of Stephen McCabe's dear friend Joseph Bourne. Mrs. McCabe, the mother of five children, was fifty-three years old and suffered from heart disease.

Two weeks later, the Bourne family left Meriden and returned to Cambridge, Massachusetts.

On June 19, the paper noted that "Chas. Monroe of the Glass Works, is developing fast into a 'champion snake killer.' Last week he killed a large water snake near the culvert between the Organ and Glass shop. His snakeship was in the act of devouring an eel, and measured three feet, four inches. This morning he spotted another large one, but while reaching for a club the snake slid out, but his doom is sealed."

On June 26, the paper reported that "the 'glass house' of the Flint Glass Works will shut down next Saturday for repairs. As the factory is pushed with orders, the repairs will be made as expeditiously as possible. It is just the weather to put out a fire. It may take six weeks to do the work."

On July 1, the *New York Telegram* printed an interesting article that stated that, out of about one hundred flint glass manufacturers at that time in the country, only four sold their goods through New York City dealers. These were the Boston and Sandwich Company, Corning Glass, C. Dorflinger & Sons and the Meriden Flint Glass Company.

The seeds of collective bargaining, having been sown four years earlier with the establishment of the AFGWU, started sending up tentative shoots by the summer of 1882, according to the July 11 *Daily Republican*:

> *The delegates' convention of glassmakers of the United States and Canada held in Brooklyn last week, was fully attended, there being 88 delegates present. There was a strong feeling among the delegates to more closely unite the interest of the employer and employee, and by another year it is hoped to have a meeting of the manufacturers and the workmen, where the mutual interests of both can be discussed. Stephen McCabe and James Moore were the delegates from this city.*

On July 26, the paper announced a tragic event:

> *SAD ACCIDENT TO CHARLES MCCABE*
>
> *The telegraph brings the sad intelligence that Charles McCabe, a son of Stephen McCabe, of this city, met with a sad accident at Boston, Tuesday, which has necessitated the amputation of his leg.*

A New York, New Haven and Hartford locomotive leaving Meriden. This photo was taken in the summer of 1882, according to the note on the back. *Author's collection.*

> *Mr. McCabe was at the Boston & Lowell railroad, and in attempting to take a train for which he was a little late, fell between the cars, the wheels passing over his leg and crushing it.*
>
> *Mr. McCabe is about 20 years of age and has been visiting South Boston, the former home of his father. His father has been with him in Boston. The sympathy of the many friends of Mr. McCabe will be extended to the doubly afflicted family, the mother having died very suddenly a few weeks since. He was employed with his father at the Glass Works.*

The August 12 *Daily Republican* noted that "the glass furnace has been heating up the past week and will be hot and ready for the glass 'blowers' to resume work Monday. The needed repairs have all been made."

On August 17, *American Pottery and Glassware Reporter* had a full front-cover advertisement:

> *Meriden Flint Glass Co.,*
> *Meriden, Conn.*
> *We desire to call the attention of the*
> *trade to our Satin Finish for cut Crystal Tableware.*
> *By this patent process of finishing, we are*
> *enabled to perfect cut glass that excels in brilliancy*
> *and finish anything in the market.*
> *Our new Fruit Bowls and Flower Troughs combined*
> *are the latest novelty.*
> *Send for Photographs.*

Two weeks after the workers returned to labor, they were locked out by management in response to a demand for increased wages.

At this time, there were 150 male and 9 female workers at the Meriden Flint Glass Company. They worked sixty hours per week, earning an average of $1.78 daily for men and $0.85 per day for women.

The lockout lasted from September 1 to October 2, after which the glasshouse was obliged to run two shifts, night and day, to keep up with orders. Wages remained unchanged.[15]

There was no mention of this action in the local newspaper. While the workers were out, Charles McCabe returned to Meriden ready to work, and Charles Monroe took the opportunity to open up a shop of his own.

On October 28, the *Daily Republican* had an advertisement in the classified section.

ART ROOMS
AND
MERIDEN FLINT GLASS CO.'S SHOW ROOMS.
Oil Paintings, Photogravures,
CutGlass and Decorated Ware,
Bric-a-Brac, Etc.
The inspection of the public is cordially invited.
Instruction given in decorating China.
C.F. Monroe,
34 West Main Street

Chapter 8

1883

On January 22, Simpson Hall & Miller, a silver company in the neighboring town of Wallingford, began manufacturing glass for its own use. It had been purchasing glass blanks from Meriden that it decorated in its own cutting department.

The name of its new concern was the Morse & Brooks Glass Company. About fifty men were employed to work the eight-pot furnace. Many of the workers came from the Meriden glassworks. Six months after opening for business, the newspaper reported that the shop had burned down but was to be rebuilt.

Back in Meriden, more flint glass workers were venturing out on their own. Glass cutters Thomas and James Niland were brothers who learned their trade in Birmingham, center of the glass industry in England. Thomas was employed at Boston and Sandwich and the New England Glass company before coming to the Meriden Flint Glass Company in 1881. Two years later, he and previously fired co-worker James D. Bergen opened their own glass cutting establishment in town. Thomas's brother, James, would arrive soon to work with them.

On March 5, the *Daily Republican* noted that "Bergen & Niland will commence work in their glass shop tomorrow."

On May 7, the paper reported that "in the game played between the Organ and Glass shop nines in North Meriden, Saturday afternoon, the former were victorious by a score of 26 to 16. Several of the Organ shop boys had their fingers split in their frantic attempts to stop the ball as it flew around the field." (Baseball was played barehanded in 1883.)

A glassworkers' cottage on Tremont Street in 2011.

The glassworks was shut down on Friday, June 14, its annual custom, to make repairs. During the break, the men played weekly games of baseball with the other factory teams.

On August 28, "the Flint Glass shop base ball players say that they have beaten the Organ works and other shop nines until they are tired, and they are eager for fresh worlds to conquer," according to the paper. That fall, a new railroad began laying tracks behind the glassworks and removed two cottages from there to Tremont Street, where they still stand.

The Wallingford glass factory was torched twice more before the end of the year. The premises had been saturated with turpentine, indicating incendiary work. The company was fully insured.

Chapter 9

1884

The Meriden glassworks operated fairly smoothly and quietly during 1884, although workers continued to leave. Early in the year, the firm's George E. Hatch, along with MFG bookkeeper Thomas B. Clark, opened a glass cutting shop in Honesdale, Pennsylvania. By March, the glass shop nine had reorganized and issued a challenge to any other shop team that cared to take it on.

In late June, when the factory shut down for the annual break, the *Daily Republican* noted that "Thomas Brooks, a Wallingford glass cutter, who formerly worked in Meriden, eloped with a Mrs. Blakeslee of that town last year, leaving a faithful wife and several children for his new love. Mrs. Brooks has been keeping track of the movements of her husband, and had finally succeeded in having the pair arrested in Pittsburg. She has gone to that city to testify against them."

Roller skating parties were a popular form of entertainment. Originally held in the city hall auditorium, they proved so successful that a new rink was built for the purpose on Hanover Street. Charles Monroe was the manager of the new facility. The city was being lit by electricity for six hours each night, and Mr. Monroe kept his art rooms open evenings to sell roller rink tickets.

Meanwhile, people who intended to place electric lights in their stores or homes were reminded that the lights would invalidate their insurance unless they received permits from their insurance company. Apparently, it was considered safer to light one's home with a kerosene lamp than with electricity.

In October, Hatch and Clark had a falling out, and Mr. Hatch withdrew from their Honesdale enterprise, which T.B. Clark continued to operate successfully for many years.

The December 19 *Daily Republican* reported that "the sociable of the Flint Glass cutters takes place, this evening, at Alfredian hall. Fleischer will furnish music, and Rogers will prompt. The committee guarantees that all who go will have a pleasant time."

Chapter 10
1885

While Meriden was building its first public high school building, another movement was underway to establish a hospital. Various fundraising events were organized to finance the venture. Donations were solicited at the various factories, and roller-skating rink parties were held for the purpose of raising capital to build the hospital. Charles Monroe, as manager of the skating rink, was an active participant in the fundraising activities. Benjamin Ray and Stephen McCabe were in charge of collecting contributions from workers at the glassworks.

By early March, however, fundraising activities at the glass factory had been curtailed by yet another strike. There was no mention of the walkout in the paper, except for one brief note. On March 11, the *Daily Republican* reported that "the difference between the Flint Glass company and the glass makers in three of their 'shops,' in regard to work, has been adjusted and the men will in all probability go to work, next week. They will go on short time rather than do extra work on full time…There is perfect harmony between the men and the company."

The "perfect harmony" was short-lived. Ten days later, thirty-one glassblowers and sixty-one of their shop mates struck "against deductions for imperfect work." The walkout, ordered by the flint glass workers union, lasted for fourteen days, but its effects were permanent. The strike was not successful, and wages remained unchanged. The firm, however, was dealt a crippling blow from which it was unable to recover. The company could get no men to fill the strikers' places and had to suspend most operations.

The May 5 *Daily Republican* noted that "William Hobson, of 49 Tremont street, and a well known glass blower, who has been employed for a long time by the Meriden Flint Glass company, died suddenly, Monday night. He went to work as usual, Monday afternoon, and without a moment's warning he fell on the floor in an apoplectic fit. He was carried home, and died Monday night. Mr. Hobson was highly esteemed by all his fellow workmen. The funeral will be held Thursday afternoon, at 2 o'clock."

Over the next few months, the Meriden Flint Glass Company hemorrhaged workers. However, the entire Eastern Association was experiencing difficulties at this time as well, and workers bounced from factory to factory. By fall, nearly every glass factory in the East had closed down, the manufacturers refusing to accede to workers' demands for better wages. On September 17, the *Daily Republican* reported that "Wm. Flannigan, formerly at the Meriden Flint Glass works, started for Chicago, Wednesday evening, to take charge of the decorating department of a firm in that city."

The next day, the paper noted that "Arthur and Charles McCabe, sons of Stephen McCabe, left this morning for Carrington, Penn., to work in the glass shop there."

William Hobson's Tremont Street home, in 2011.

On October 9, the paper reported:

> *Manufacturers of flint glassware in New York, Brooklyn, Philadelphia, Boston, New Bedford, Mass., and Meriden, met quietly at the Astor house, in New York city Wednesday, and unanimously determined to refuse the demands of employes* [sic] *for increased wages and a decreased output. Another meeting was arranged for in about ten days, when the employes have acted on this. There is now an extensive strike in the glassware works. All the flint glass factories in the eastern association, comprising Pennsylvania, NY and New England, are idle, save only the Boston & Sandwich works, the Mt. Washington factory at New Bedford, and the New England factory at East Cambridge. This includes factories in Brooklyn, NY, one in Philadelphia, one in Meriden and one in Somerville.*

With his former employer out of operation, James Bergen's glass cutting establishment stepped up production to satisfy the Meriden Britannia monster's voracious appetite for caster bottles, previously the bread and butter of the flint glass works. Eydam & Handel continued doing handsome work in glass decorating, Charles Monroe's shop kept him busy and two former MFG decorators, Emil Kroeber and Albert Kaeppel, went into the glass decorating business, taking quarters in a separate building right on the factory grounds.

Chapter 11

1886

On January 19, the *Daily Republican* reported that "the stockholders of the Glass Works held a meeting this afternoon, at 2 o'clock, but adjourned without doing anything. The date of the adjourned meeting has not yet been decided on."

The April 22 *Daily Republican* noted the following:

> *George E. Hatch, formerly secretary and agent of the Meriden Flint Glass company, is about to go into the glass cutting business in Brooklyn, N.Y.…Mr. Hatch was here to-day and made Daniel Forbes a flattering offer to take charge of the cutting shop. Mr. Forbes was foreman at the glass shop here and since business was suspended at that shop he had been running a small cutting shop on South Colony street. He is considering Mr. Hatch's offer and the only thing that deters him from accepting it is his dislike to leaving Meriden, where he is comfortably located.*

On April 26, the *Meriden Daily Journal* reported that "Daniel Forbes has concluded to leave Meriden and work for Hibbler & Rausch glass manufacturers, in Brooklyn, N.Y. This is the concern in which George E. Hatch is interested. 'I don't like to leave Meriden,' Mr. Forbes said to-day, 'but the offer I have had is too good to refuse. In addition to being paid for all I have expended in fitting up my South Colony street shop, I have a contract for two years at a salary of thirty dollars a week.'"

The May 6 *Daily Journal* noted that "Frank Morse of Wallingford was arrested to-day on the charge of burning the Morse & [Brooks] glass shop in Wallingford in 1883. He was bound over under $8,000 bonds for burning the shop and under $1,000 for defrauding the insurance companies. Bonds were furnished by Mr. Morse's father."

While a steady exodus of former employees headed to the glass factories of Ohio and Pennsylvania, it appears that Meriden's furnace itself was being kept in limited operation by the Meriden Britannia Company, continuing to produce a modest number of glass blanks for decoration. This is supported by the fact that Kroeber and Kaeppel had opened their glass decorating shop on the factory grounds. Further, there were thirty people listed in the 1886 Meriden City Directory as being either pot-makers or glassblowers. The only place they could have been working at these trades would have been at the glass factory.

Meanwhile, the other vacant factory buildings on the premises were being trotted out before a variety of prospective tenants, such as a Boston harness-maker, the Norfolk Axle Works and an organ manufacturer out of New York City.

The August 3 *Daily Republican* reported that

> *the Meriden Flint Glass Co.'s factory has been leased permanently for the Mechanical Orguinette Co., except the furnace rooms, which will not be disturbed, as they in all probability will be needed for glassmaking in the near future. The factory will be connected with the Wilcox & White Organ Works by a bridge, to better facilitate shipping &c.* [See March 13, 1888.]
>
> *The Orguinette Co. make the orguinette in all designs and sizes, from the miniature instrument to the most beautiful organ with mechanical attachments costing $1,000. The instruments are very popular, as they play mechanically, and render the most beautiful music. Meriden is indebted to Mr. White for bringing us such a solid concern.*

The August 14 *Daily Journal* reported the following:

> *The Long Strike*
> *Men Who Helped to Close the Flint Glass Works Leave Meriden*
>
> *George Franklin, James Conway and Frank Corey, three glass makers, to-day left Meriden for Cumberland, Md., where they have secured*

> *work through the efforts of the Flint Glass Blowers' association, of which they are members.*
>
> *It was sixteen months ago that thirty-one men employed at the Flint Glass works here went out on a strike. The company could get no men to fill their places and eventually had to shut down and now the shop is for rent, the glass business being practically abandoned.*
>
> *Of those thirty-one men all but six have now been found work by the association to which they belonged. All the time the men have been idle they have received from the association six dollars each a week, which in the case of the three men who left to-day would make about $400 per man. It was a long fight, but the men were determined and moreover, were backed by their powerful association. It is expected that soon the men not yet at work will be sent where they can find plenty to do.*

The September 8 *Daily Republican* noted that "C.F. Monroe, the glass man, has secured the services of Karl Helmschmied, a first class artist. The man comes from Steinschonam, Bohemia, and his work is really of a superior character, being full of design and excellent workmanship."

The *Daily Journal* on September 30 reported that "Thomas Franklin, glass maker, left for Pittsburg, Pa., last night, where he will work. Henry Cook is now the only one of the men in the employ of the Flint Glass company who went on a strike some months ago that has not found work outside of Meriden."

By the end of the year, Stephen McCabe announced that he was opening a bookstore at 82 East Main Street. While specializing in Catholic books and goods, he would also carry a full line of toys, confectionery and stationery, as well as standard works of the leading poets and other authors.

Chapter 12

1887

In April, the LaBelle Glass Company of Bridgeport, Ohio, made Stephen McCabe a very flattering offer to act as manager of its glassworks, with a tempting salary. Mr. McCabe hesitated, as he had just gotten his Meriden bookstore on a paying basis, but in late April, he decided to go out and "take a look." This was no doubt influenced by the fact that two of his sons were employed there. In the meantime, his daughter, Louisa, who had been his saleswoman since the opening of the bookstore, would continue to conduct the business.

Mr. McCabe ultimately decided not to accept the LaBelle offer. He found the glassworks' legal affairs to be in "a muddled state." Furthermore, the natural gas in general use there left "a grimy, oily substance…which makes life miserable, and a clean collar or pair of cuffs as rare as a day in June."

One year earlier, a number of flint glass manufacturers had sent a letter to the president of the AFGWU. The letter stated that the manufacturers had appointed a committee that desired to meet with a like committee from the union to discuss and, if possible, agree on move lists and rules for the ensuing blast. These committees met in July 1887 and agreed on lists and rules for the 1888 blast.

About 125 years later, the American Flint Glass Workers Union had more than twelve thousand active members. In 2003, a tentative merger agreement was announced between the AFGWU and the United Steelworkers of America, making it into one of the largest of numerous modern unions to represent glassworkers in the nation.

On July 2, the *Meriden Daily Journal* reported that "the orguinette folks will have to vacate their present quarters as the new harness concern from Boston has hired it. The plan now is to unite the present Orguinette company, which came here from Brooklyn, with a company in Boston that makes the orguinette cases. The projectors wish…[to] build a large shop between the Wilcox & White Organ shop and the Glass shop."

In July, a U.S. Labor Bureau agent arrived to obtain information about the March 1886 strike at the Meriden Flint Glass Company. The local paper speculated that the glassmakers' union, thinking that it had been paying benefits long enough, sent the agent to gain what facts he could in order to make a settlement with the strikers. The agent also inquired as to what had led to the suspension of business in the first place.

At this time, George Wilcox verified a rumor that the glass factory was to resume operation, once the Orguinette Company had moved into its new building under construction on Tremont Street. Glassmaker James J. Murray of Philadelphia had agreed to rent the facility.

On July 16, the *Hull Village News* of Nantasket Beach, Massachusetts, published the following informative snippet about yachting enthusiast Charles F. Monroe, whose company would later produce a line of decorated opal ware known as "Wave Crest": "Messrs. Arthur Brooks, C.F. Monroe and E.H. White of Meriden, Conn., have been cruising about the bay for a couple of weeks in the sloop yacht Wave Crest…anchoring here on account of fog. They had a narrow escape on their way to Provincetown, a sudden squall nearly capsizing them. Their first night's experience was too serious for pleasure. The girls at home would hardly recognize these 'tony society gentlemen' in the bronzed, blue-shirted yachtmen."

The news that the glassworks was to start up again brought out a curious item. The July 23 *Daily Journal* reported the following:

> *Apropos of the small death rate of Meriden and as a matter of cheerful contemplation during the present cholera producing weather* THE JOURNAL *rises to remark that there is one thing that is generally unknown to the perspiring citizens, and that is that a number of well known capitalists, whose belief in the efficacy of cremation is as strong as their financial legs, have had under contemplation the formation of a stock company for the purpose of buying the old glass works and establishing a furnace there that would draw trade from the surrounding vicinity and while not perhaps being a paying investment would materially advance their belief that cremation is the only proper way to dispose of the dead.*

> *The matter has been worked up quietly with the intention of ultimately getting hold of the glass works as the only available structure unless one was built and that would cost too much. The news that the glass works were about to start up again was conveyed to one of the protectors of the enterprise who is an enthusiastic cremationist and he sadly remarked that the scheme was at an end then.*
>
> *"Our plans," said he "were nearly ready to put into practice but of course we can't do anything now. It would cost a good deal of money to build the proper kind of furnace, a good deal more than the new company could pay, whereas the furnaces at the glass works are peculiarly fitted for the purpose we need, and for only a slight outlay we could have had one of the finest crematories in the country. Of course we didn't expect to make money at first; that wasn't our object but we hoped to educate the people to the fact that crematories are what are needed in the country."*
>
> *He grew enthusiastic as he dilated on the virtues and rapid growth of cremation. "Don't mention my name for the world," he said, "for I have another scheme that may work yet, but if any one wants to know anything about it send them to me quietly."*

On July 28, the *Daily Republican* noted:

> *The new proprietor, Mr. Murray, has everything in his favor for the building up of a successful business. Lamp chimneys will probably be a staple article of manufacture, and Mr. Murray has ample capital to back up his venture. The advantage of Meriden's location is greatly in advance of that of the Western manufacturers. The sand needed in the manufacture of the cheaper grades of glass is found in immense quantities near New London. The ordinary sand hereabouts is too flinty. A soft, silicious sand is necessary. Seven grades are also found at Berkshire, Mass., the finer ones being exported to foreign countries for use in glass making. The freight rates to New York are of course, also, much cheaper than from the great Western factories and much loss from breakage will be done away with by the shorter haul. Taking everything into consideration the outlook for the new enterprise is remarkably bright, and the 150 hands who will be employed will add a good element to the city's population.*

The August 1 *Daily Journal* noted the following:

> *For the first time in over two years the fire in the big furnace at the Flint Glass shop was started Saturday night, it being the first move in the work of resuming the manufacture of glass in this city...It will take over a fortnight for the fires to become in condition to make glass.*
>
> *As the smoke curled up yesterday from the big stack the people of the North end looked pleased at the prospect of the new boom which the Glass shop, with the Orguinette and Harness factories will give that section.*

James Murray & Co. had a rough time starting up. Local landlords, anticipating the arrival of new tenants, raised neighborhood rents, and the glassworkers found it difficult to secure lodgings when they arrived in town.

Two days after production began in mid-August, half of the men had to stop work on account of the heat.

Two weeks later, a discharged employee became violent and attacked Mr. Murray with an iron pipe.

In September, a number of glassworkers were accused of assaulting a streetcar driver.

During the first month of operation, those workers who had been out of work for a while required tannic acid to toughen their hands, which had become tender while unemployed at their craft.

By mid-October, half of the crucibles had broken and the entire factory had to be shut down for a week while the pots were replaced and the furnace brought back up to heat.

At the end of October, the bookkeeper quit and moved to New Jersey.

In early November, the factory had to stop production because there was "no glass plain."

The crude petroleum being burned in the furnace sent up such a stench throughout the neighborhood that a petition was circulated asking that the nuisance be abated. Among the signers was W.W. Lyman.

By the end of November the factory still did not have enough pots and was not yet running full time.

Finally, in December, the entire eastern association of glassmakers, which included Meriden, was ordered out on strike by the union.

The *Daily Journal* of August 31 reported that "the glass works have been running nights and business is just booming. Kroeber & Kaeppel, who buy glassware from J.J. Murray & Co. for decorative purposes, pronounced the

View of the glassworks from Cambridge Street, circa 1886, showing Kroeber & Kaeppel's decorating shop. *Author's collection.*

goods first class in every respect and not surpassed by the product of any other glass shop in the country."

The October 18 *Daily Republican* noted that "the large number of Glass works people on the street to-day is accounted for by the fact that five, or half, of the large pots used for melting the metal have been broken in the past few days, and the entire factory has stopped work in consequence. New pots are being put in, but the shop will not run over half time this week, owing to the necessary cooling of the furnaces to put in the new pots."

The *Daily Republican* of November 2 reported that "the elegant silver trowel used by Bishop McMahon at the laying of the corner stone of the Church of the Holy Angels in South Meriden last April is on exhibition in the window of McCabe's book store on East Main street. It will be drawn for at the fair."

On November 26, the *Daily Republican* noted that "the glass factory is increasing its capacity by putting in more pots. This will give the men a chance to make full time, which they have not been able to do heretofore."

On December 1, a combination known as the Associated Glass Manufacturers of the United States was formed to bargain with the AFGWU. A lengthy scale of moves was presented to the trade union, and an executive committee was empowered to import foreign glass in the event of a strike.

Ten days later, the union ordered flint glass workers to cease work until a list of moves, wages and rules could be agreed on by the manufacturers and workmen. Thousands of glassworkers were affected.

The strikers eventually withdrew their demand for a raise in wages and conceded the right of employers to discharge workmen for incompetency and drunkenness. They proposed a fifty-five-hour workweek for moldmakers and a fifty-eight-hour workweek for cutters and refused to agree to manufacturers hiring nonunion workers.

This home was erected by Mr. Lyman in 1887 for E.A. Chapman, an employee of the new Orguinette Company. *MHS.*

The Chapman house in 2011. Joseph Bourne's former residence is visible to the right. Both are directly across the road from Mr. Lyman's "new" house.

The *Daily Republican* of December 28 reported that "a petition is being circulated among the residents of the north end, and signed by W.W. Lyman, E.A. Chapman and others asking that the nuisance, which results from the bad odor of the crude petroleum oil used in reheating the furnace at the Glass works, be abated. The smell is very offensive to those living in the neighborhood of the shop, and the company will, no doubt, contrive some means of having the atmosphere somewhat more agreeable when it is necessary to reheat the furnace."

CHAPTER 13

1888

In early January, two Pittsburgh glass manufacturers took the decisive step of leaving the Glassmaker's Alliance in order to reopen their shops. On January 9, the Murray glassworks in Meriden did the same.

The January 12 *Meriden Daily Journal* reported that

> *the flint glass blowers are jubilant over the news that James Murray's glass factory has withdrawn from the combination of manufacturers, and that Mr. Murray has informed his men that they could return to work at the old terms. Mr. Murray has written to all the eastern firms that he will ship goods throughout the country by Saturday, and under no conditions whatever will he join the Manufacturers' association again. This, the strikers say, breaks the combination, and will compel the other manufacturers to give in.*

Not everyone was jubilant over the news, though. Pennsylvania glassmaker Christian Doerflinger came to Meriden to confer with the renegade glassmaker about his departure from the union.

The January 16 *Daily Republican* reported that "the Glass Workers' union held a meeting Saturday afternoon, but no action was taken on account of the troubles between the manufacturers. It is said the meeting was called on account of the presence in the city of Mr. Doerflinger of White Mills, Pa., who is not liked by many if not all of the members of the union."

The *Daily Republican* of March 13 carried the following story:

> *In the evening of Monday the city presented an awful and picturesque appearance to those whose duty called them out…There were no lights burning, and the weary wayfarer was obliged to pick his path among the enormous drifts in Egyptian darkness. The main streets were illumined to a certain extent by the lights in the store windows, which the merchants, in spite of the fact there was no trade, in most instances kept burning brightly. But the rays of light in many cases didn't extend further than to the sidewalk, because of the drifts which in many cases extended up to the tops of the show windows, 20 feet above the road bed. There was no attempt at travel by teams.*

The next day, the paper reported that

> *the Glass shop door on the east side is barricaded by a bank fifteen feet high. The bridge connecting the Glass and Organ shops is a total wreck, smashed to smithereens by a big snowflake that struck it Monday night. Two men had but just stepped off from it as it fell; a narrow escape.*
>
> *Fifteen men and boys were imprisoned in the Glass shop during Monday night and until Tuesday afternoon by the storm. It was almost sure death to venture a homeward trip. Two good and true men made a venturesome trip to Larry Geisler's grocery and returned with crackers and cheese for the famished ones. The wind blew a perfect gale, the snow being so blinding that a hand had to be held close to the face to see it.*
>
> *Tuesday afternoon a few Norwegians ventured forth clad in show-shoes made from barrel staves. The men from the Glass shop also got out by the same device. It was fun to see some Swedes perambulating the snow drifts at the same time-their locomotion was after the manner of the bird called a porpoise. It was a "roll and go."*

For as long as the Eastern Association strike continued, each man who had work, including those at Meriden, had to pay two dollars a week to support the strikers.

By April, a number of Murray glassworkers announced their intention to return to Philadelphia. Perhaps they knew something?

The May 1 *Daily Republican* noted:

> [Kroeber & Kaeppel], *a lamp shade decorating firm, has its shop opposite the Murray Glass Works. Most of the shades are purchased from the glass works, decorated, and then sold to dealers. This morning it was*

> *discovered that there had been a robbery in the shop. Entrance had been effected by breaking a window and 18 shades, valued at anywhere from $75 to $100 had been stolen.*
>
> *Suspicion fell on Wm. Burke, a glass blower, and his room at his boarding house on Pratt street was searched, and five of the stolen shades were brought to light. Burke was placed under arrest.*

Edward Drummond Libbey, son of William L. Libbey, had taken over the New England Glass Company after his father passed away in 1883. Libbey the younger steered the Cambridge factory through numerous financial crises and strikes. After five years of unsettled conditions, he famously told his men that if one more strike occurred, he would shut down the works and move the operation west. He was a man of his word.

The May 4 *Daily Republican* reported that "E.D. Libby, proprietor of the New England Glass Works at Cambridge, Mass., who is now erecting one of the largest glass factories in the west, is in the city, and has made flattering proposals to Stephen McCabe to accept a prominent position in the new works. Mr. McCabe will not accept, his business interests in this city requiring his entire attention. He…has bought out the undertaking business lately conducted by R.J. Reilly, and will conduct the same in connection with his bookstore at 82 East Main Street."

The *Daily Journal* of May 12 noted the following:

> *C.F. Monroe has bought the lot west of the Meriden & Waterbury railroad track on Capitol avenue and will build a factory there, in which to carry on his increasing decorating business.*
>
> *The shop will be seventy-five feet long, thirty-five feet wide and two and a half stories high. Mr. Monroe will put in four new kilns in addition to those now in use at his present shop and employ twenty-five new hands.*

Mr. Monroe operated his business at that location until 1916, producing Wave Crest, Kelva and Nakara decorated opal wares and, later, cut crystal.

The May 14 *Daily Journal* noted that "the glass makers and cutters of the Murray Shop will play a game of ball next Saturday on the grounds near Gehrmann's grove for a purse of $25."

Gehrmann's Grove was located on Broad Street, just north of the city line. The ball field was on the other side of the new tracks that ran behind the glassworks.

C.F. Monroe's Opal Glass Decorating Works on West Main Street, Meriden, Connecticut. *Author's collection.*

The May 21 *Daily Republican* reported that "the glassmakers had no quorum at their Saturday afternoon meeting, owing to the ball game reported elsewhere." The next day the paper noted that

> *there are rumors afloat that Meriden may lose her infant industry, that of glass making, but they lack verification, and, in fact, are denied by some who ought to know...The rumor says that Mr. Murray may build or buy a shop near Philadelphia, which is his home, and transfer his business to that place, meanwhile holding his present lease of three years on the factory here...Mr. Murray is reported to have denied the rumor, but the men still believe that some such arrangement may be made. Meriden would not like to lose the glass industry from its growing manufacturing interests.*

Mr. Murray closed the Meriden operation on June 23. When asked why, he replied that it was too expensive to do business here. His freight bills were enormously high, and he was simply losing money by remaining in town. Within days, the tools and fixtures at the glassworks were packed and removed to Philadelphia, and nearly all the glassmakers had left town.

The July 2 *Daily Journal* noted that "Thomas Niland of the State street glass shop went to New York this afternoon and to-morrow will sail for

Europe…to buy glass." The same day, the *Daily Republican* reported that "Arthur McCabe left today for New York, where he will spend about two weeks, and from there he will go to Philadelphia to join the J.J. Murray glass forces."

The July 23 *Daily Republican* reported that "a slight fire occurred in Kroeber & Kaeppel's glass decorating shop, early this morning, and a still alarm was sent out. The fire did not amount to much. It was extinguished before much headway had been gained. A hole, about a foot in diameter, in the floor of the drying room, was the total damage. The members of the Byxbee Hose company who sleep in the hose house went up there with Chief Horan, but didn't use much water. The cause is unknown."

The *Daily Journal* of August 7 noted the following:

> *Meriden people will be pleased to learn that the probabilities are strongly in favor of the Glass works again being started up. Rumor has it that the Ravenswood Glass company of Long Island are to locate here and as a verification of this it is known that the proprietors of that concern have been here and looked over the works recently vacated by J.J. Murray… The gentlemen interested with the Meriden men who own the works have looked over the ground and it is said are so well pleased with the appearance of things that they will come here.*
>
> *The great glass works of E.D. Libby, which have but recently removed from Cambridge, Mass., to Toledo, Ohio, will employ several of the old Meriden glass makers. Among them will be C.S. McCabe, son of Stephen McCabe of this city, who will start for there on Friday.*

The August 9 *Daily Republican* reported that "as far as can be ascertained there is no truth in the rumor that the Ravenswood Glass company of Long Island, is to start up the Meriden Glass works. About two months ago a representative of that concern was here and looked over the idle works, but no word has been received from him or the company, and there is no expectation that anything will be done."

The *Daily Journal* of September 26 noted the following:

> *Ironworkers, glassworkers and others are constantly exposed to a heat of 400 or 500 degs. F., and yet do not become burned, and there can be little doubt that the enormous radiation from heated rocks and valleys, in addition to the direct rays of the sun, make up an amount of heat far greater than is ever experienced on even a very sunny snow slope, and*

yet one does not become sunburned. No doubt the surface of the snow reflects and disburses much heat, but certainly far less than it receives, as heat rays are absorbed and rendered latent by the snow melting and evaporation. Experience fully corroborates this, for one may lie on one's back and freely expose the face for long periods to the sun and yet remain unburned. There must, therefore, be some one factor in sunburn than heat alone.

Chapter 14

1889

The March 23 *Daily Journal* reported the following:

> *Fire broke out in the glass decorating shop of Kroeber & Kaeppel, up by the glass works, about 5:30 last evening. As the workmen were working on the bell at the time the alarm was not struck at first and it had to be made with a hammer. There were two alarms on this account. The building in which the fire took place was a two story wooden one about 50x20 and was connected with the main building of the glass factories by an iron foot bridge from the second story.*
>
> *The cause was an over heated stove it is supposed, as no other reason can be assigned. The men quit work at 5 o'clock, so the building was vacated at the time. It was completely gutted and the stock was a wholesale ruin by the time the firemen were through. Being of glass mostly it can easily be seen how it must have suffered.*

The *Daily Journal* of April 4 reported the following:

> *Dennis McCarthy created a sensation and almost died in Stephen McCabe's bookstore and undertaking establishment. Shortly after 7 o'clock he went to the store in rather a dilapidated condition, evidently under the influence of liquor. After talking with Mr. McCabe, whose assistant he is, for a while, he said, "Good-bye, McCabe," and seizing a bottle of embalming fluid drank nearly a pint before Mr. McCabe had a chance to prevent him.*

The bottle was dashed from his hands. The principle ingredient in the mixture is arsenic and it quickly took effect. McCarthy collapsed under the powerful influence of the drug and was laid out on some boards on the floor. Dr. Hughes and the young man's father were immediately sent for. After three hours' hard work the physician by means of emetics and a stomach pump managed to draw the poison before it had completed its work, but McCarthy was left in an exhausted condition. The chances are that he will come out all right. The act was apparently done under the influence of liquor. McCarthy's sister died only last week.

The June 18 *Daily Journal* reported on the following:

The news of the death of Stephen McCabe yesterday afternoon was received here last night and created great surprise and much sorrow. Mr. McCabe left here last Thursday, with his daughter Louisa, to spend several weeks with relatives near Boston, where he formerly lived, in hopes of benefiting his health, which has been in a precarious condition for several weeks.

Miss McCabe returned Saturday to look after Mr. McCabe's business on East Main street. She left her father in Boston feeling quite well and confident that a few weeks' rest among his friends would result in greatly benefiting him. The telegram announcing his sudden death was, therefore, a great surprise to Miss McCabe. The dispatch gave no particulars, but is understood that death was caused by heart disease Mr. McCabe was a very large and fleshy man and always feared that his death would be sudden.

McCabe came to Meriden to work for the old Flint Glass company several years ago and was one of the principal foremen of the factory. He thoroughly understood the glass making business and had worked in several of the principal shops in the country as foreman and superintendent.

About three years ago he opened a book and news store in Morse & Norton's block. He added undertaking to the business some time ago and was doing a snug business until his health broke down and a nervous attack unfitted him for trade.

The funeral will take place at Boston, where his wife, who died in this city, is buried. Mr. McCabe's death will be mourned by a large number of friends, several of whom went to Boston to-day to attend the funeral.

Stephen McCabe was fifty-six years old. A Civil War veteran, he had been shot in both legs during the Battle of Chantilly in 1862. He left behind his daughter, Louisa, and three sons, all glassblowers.

The *Daily Republican* of July 13 noted the following:

> *Within the past few years the "north end" of Meriden has been undergoing a transformation, and to-day it is one of the busiest sections of the city. The manufacturing concerns that have located there within the past few years seem to be in as prosperous a condition as those located nearer the center of the city.*
>
> *At the Wilcox and White Organ company's plant another illustration of how this part of the town is booming is noticeable. The Organ company and the Meriden Flint Glass Works located here almost simultaneously, and the stock of the Glass company was generally preferred among the capitalists of the city. Yet while the Glass company is defunct, the Wilcox and White Organ company has constantly added to its original plant and at the present time has an office undergoing construction that when finished will be second to none in the city.*
>
> *Standing close to this is another manufacturing concern, the Aeolian Organ and Music company, which has been with us but about three years. This company is the result of the consolidation of the Mechanical Orguinette company of Greenpoint, N.Y., and the Automatic Music company of Boston. The first named company located here first, occupying the old glass house for the construction of its instruments. Later the two*

The Wilcox & White Organ Factory. The glassworks chimneys can be seen in the background. *MHS.*

The Aeolian Organ and Music Company, 2011.

> *companies were reorganized under the title the company now bears, and the elegant factory was erected at the north end. Success seems to have crowned the efforts of this concern and it contemplates building more extensively.*

Reporting on a gala event, the December 21 *Daily Republican* noted the following about W.W. Lyman's granddaughter, whose ninth birthday party had been so elegantly celebrated nine years earlier in her grandfather's new home:

> *Miss Etta Warren returned home from New York, Friday, and brought with her, as her guest through the holiday vacation, Miss Spofford, a pupil at Mrs. Sylvanus Reed's school. An impromptu party was tendered to a few friends at Miss Warren's home in the evening. Music and dancing helped to pass away a most enjoyable evening. A light luncheon was served. Among those who attended were: The Misses Spofford, Fanny Hall, Emma Thomas, Etta Warren, and Messrs. Stillman and Benjamin of Hartford, and Walter Coe, Arthur Miller and H.M. Billard of this city.*

Chapter 15

1890

On March 24, the *Daily Republican* reported that "Kroeber & Kaeppel, successors to the Meriden Flint Glass company in decorated glassware, have sold their fixtures, materials, etc., to Eydam & Handel, who will supply their customers in the future in connection with their own wares. Messrs. Eydam & Handel are popular young men, and they will, no doubt, increase their business more than two fold on account of the acquisition of such a competent firm."

The June 28 *Daily Republican* noted the following:

> *Through the instrumentality of the Meriden Britannia company Meriden has secured a new industry which will benefit the north end particularly and the entire city generally. The E.A. Bliss Co. of North Attleboro, Mass., manufacturers of novelties and silver-plated goods, is the concern in question. Negotiations for the transfer of the company to this city have been conducted since last January.*
>
> *The concern has leased the old glass factory for a term of five and one-half years. The company has a capital of $50,000, employs 100 hands, and has been in business for the past eight years.*
>
> *Mr. E.A. Bliss arrived in Meriden at 2:57 this afternoon, for the purpose of visiting the glass factory to see what changes will have to be made in the plan of the building, which will be remodeled to suit the convenience of the business. Work on the building will be commenced at once and pushed rapidly, as it is expected to commence work in Meriden August 1.*

Britannia Street trolley, circa 1890. *MHS.*

The firm manufactured, among other things: calling and playing card cases, pocket match safes, bon bon boxes, stamp boxes, pocket combs, knives, button hooks, glove buttoners, paper cutters, mail openers, vinaigrettes, odor and liquor flasks, dress buckles, garter buckles, chatelaines, shopping and memorandum tablets, chatelaine bonbonnieres, specie purses, powder boxes, spectacle cases, pen wipers, paper weights, pocket mirror cases, Roman and gold plated bead necklaces, gentlemen's suspender sets, cigar cutter charms.

Chapter 16

1891

In 1891, Eydam and Handel dissolved their partnership. Mr. Eydam went to work at C.F. Monroe's, while Mr. Handel continued the business at the East Main Street building.

Glass cutter James Niland married Miss Emma Bergen, sister of James D. Bergen.

The *Daily Journal* of November 2 reported the following:

> *People in this section cannot but recall the once large and prosperous business once carried on by the Flint Glass company here, as well as the more recent second attempt to make it a success, all of which unfortunately was a failure in its way. With the exception of that portion of the factory now occupied by the E.A. Bliss company, which, by the way, is one of the neatest kept concerns in town, the building is rapidly losing its value, principally, of course, on account of non-use. This is the extreme eastern wing, containing in part the immense blast furnace used years ago.*

The November 10 *Daily Journal* noted that "it is a long ways from being an agreeable task, to note the fact that W.W. Lyman is rapidly failing and his long list of friends will hear of it with sincere regret, as well as also of the fact that Mrs. Lyman is confined to her bed."

William Worcester Lyman passed away on Sunday, November 15, at his residence, 155 Britannia Street. He was seventy years old.

CHAPTER 17

1892

On April 4, the *Daily Journal* reported on a freak accident:

> *During the thunder storm early this morning the lightning struck the big chimney of the old glass house at the Bliss company works and tore a jagged hole in it about twenty feet long and from one to two in width. This is the second time that it has been struck, and it looks as if it would be cheaper to put up a new chimney than to repair the old.*
>
> *Not far from the chimney is a cottage that got pretty well shaken up. The people who live there thought that it was coming down. One woman was thrown out of bed by the shock.*

The May 19 *Daily Republican* reported the following:

> *Mrs. William W. Lyman, aged 74 years and 23 days, died at her residence, 155 Britannia street, after being in feeble health for several months, at 8 o'clock this morning. Death was expected. The death of her husband, last November, was a severe shock to her, as she was greatly attached to him.*
>
> *Mrs. Lyman's maiden name was Roxanna G. Frary. She was born in this city April 27, 1818, daughter of James Frary. Deceased was married to W.W. Lyman September 5, 1844. She leaves one daughter, Mrs. Henry Warren, a granddaughter, Miss Etta Warren, and one sister, Mrs. Luther Boardman of East Haddam.*

Above: Meriden Flint Glass Company from the rear, showing the curved roof of the furnace building and the cone-shaped chimney. *Author's collection.*

Below: The Lyman house in 1978, as Crosson Convalescent Home. *Author's collection.*

Etta Warren would marry three times. She lived briefly in New York with her first husband and otherwise remained in the Britannia Street mansion until her death in 1930. She had one child, a daughter by her second husband, who inherited the home and sold it to a convalescent hospital at the time of her marriage in 1930 to a New York businessman. The house is used as a multiunit apartment building today (2011).

The *Daily Republican* of July 12 noted that "the fifteenth annual international convention of the American Flint Glass Workers' association is in session in this city."

The July 28 *Daily Republican* reported that "James Kane & son, who are to rebuild the old glass shop chimney…blew it up by dynamite and the whole mass of brick and mortar went up five feet and landed on one side. The sight is said to have been a fine one, as the whole mass went together."

The chimney was never rebuilt.

Epilogue

Some interesting events of the 1880s and the Gay Nineties as related by Russell White, a resident who lived at the North End during those years:

> *The ruins of the old glass works were a great attraction to many of the children of the neighborhood and exploring parties were often in order. On one of these expeditions several underground vaults were discovered. These*

Postcard showing the glassworks buildings in 1956, as the Napier Company. *Author's collection.*

were further explored by the aid of candlelight and were reminiscent of Tom Sawyer's "Number Two under the Cross." One of these bricked-over ovens was especially attractive as it had a small opening about two feet high, made of brick. The interior was covered with ashes and made a lovely hideaway for a certain group of boys. Old carpets, boxes of candles and matches, etc., were installed, and on many a rainy Saturday this retreat was a fine place to go to and enjoy lunch.

Later on, this was all filled in, when the E.A. Bliss Co. took over, which firm was succeeded by the present Napier Co. Possibly in a few hundred years some archaeological expedition may unearth these old ruins and become really excited over finding the many fragments of colored glass which abound there.

Meriden Glass Companies

Before the construction of the Meriden Flint Glass Company in 1876, there had been two glass decorating establishments in town, each of which purchased glass blanks from outside firms to be cut and etched by their own staffs.

The Wilcox Silver Plate Company, established in 1865, operated its own glass decorating department, as did the Parker & Casper Company, which was established in 1867 and was absorbed as a department of the Meriden Silver Plate Company in 1869, which continued to operate the glass cutting branch.

After the Meriden Flint Glass Company arrived, numerous glass decorating establishments began to flourish in town. Shops owned and operated by former glassworks employees purchased glass blanks and decorated them by painting, cutting or etching. Several of these glass cutting shops were later assimilated by local silver firms that, in turn, merged to form the International Silver Company in 1898.

Following is an abbreviated timeline of local companies:

1882–1916 *CF Monroe added a cut glass department in 1903.*
1882–85 *Bergen and Niland, glass cutters.*
1885–1922 *the J.D. Bergen Company, cut glass.*
1885–90 *Kroeber & Kaeppel, glass decorators.*
1885–91 *Eydam & Handel, glass decorators.*
1891–1936 *Handel & Company.*

1894–96	*T. Niland Cut Glass Company.*
1895	*The Meriden Cut Glass Company was established as a unit of Meriden Silver Plate Company.*
1898	*International Silver Company, which operated a number of glass decorating departments.*
1899–1917	*A.J. Hall & Co., glass decorators.*
1902–59	*J.J. Niland & Co., cut glass.*
1903–39	*Karl Helmschmied left C.F. Monroe's and opened his own glass decorating business.*
1905–84	*Silver City Cut Glass Company.*
1907–9	*Aetna Cut Glass.*
1910–23	*Webster & Briggman cut glass.*
1911–19	*H.C. Baldwin Cut Glass Company.*

In 1917, former MFG glass cutter James Callahan opened a glass decorating shop with Arthur J. Hall. Callahan was married to the sister of James and Thomas Niland. In 1960, daughter Mollie Callahan Nolan initiated an inquiry into the Meriden Flint Glass Company, where her father had once worked.

Ten years later, her research was expanded on by Lionel DeRagon, a West Hartford collector of Victorian-era glass biscuit jars. Wondering why some of his glass pieces had the Meriden Britannia logo fired onto them, he contacted International Silver Company archivist Ed Hogan, who

Above, left: Copper-wheel engraved flint "nappie" with gilt rim, made at MFG. Owned by Mollie Callahan Nolan, now in the MHS collection.

Above, right: Bottom of opal fruit bowl with decorations matching the pink biscuit jar, showing the Meriden Britannia logo fired onto the glass. *Author's collection.*

Looking up Griswold Street toward Mr. Lyman's first home on Britannia Street in 2011. To the left is a home built by Mr. Lyman for Edwin G. Pepper, an engineer at the glassworks.

The Lymans' "new" house in 2011.

The Meriden Flint Glass Company buildings as they look today (2011).

had started working for that company in 1916. The two men spent years researching the Meriden Flint Glass Company.

In 1998, I stumbled upon a 1976 article written by Mr. Hogan about the Meriden Flint Glass Company in the *Spinning Wheel* antiques magazine. Intrigued, I delved further and started putting some of my notes online. Mr. DeRagon contacted me, thrilled that someone else was researching the company. Shortly before he passed away, Mr. DeRagon most graciously gave me his extensive research materials.

This book is the result.

List of Employees

Many of these names came out of Meriden city directories, and following are the job abbreviations used in the directories. Some were listed specifically as glass cutters, glass engravers, glassblowers or, generically, as "employee" or simply "glass worker." The city directories often spelled the names differently from year to year, hence the variety of spellings in this listing. Other sources of names include Mr. Bourne's journals, where he often referred to his employees without mentioning their specific jobs, or newspapers and trade journals. Many of these people also worked for other glass shops in town or opened their own shops, but at some point they all worked at the Meriden Flint Glass Company in one capacity or another.

Legend

bkkpr = bookkeeper
emp = employee / employed at
gb = glassblower
gc = glass cutter
ge = glass engraver
gm = glass maker
gw = glass worker
JJM = James J. Murray Company
MFG = Meriden Flint Glass Company
M.B. Co. = Meriden Britannia Company

A

Adkins
Alexander, Robert, gc
Anderson, James D.
Anderson, John
Anderson, Samuel
Anderson, Wilbur/William

B

Bacon, Frank A., gc,
Bahr, Augus
Baker, Edward L., pattern-maker glass shop
Balmy, Joseph, emp MFG
Barrenthaler, William, gc, M.B. Co.
Barron, Augustine, ge
Barry/Bary, Andrew, emp MFG
Batsen/Batson/Basten, William, packer, MFG
Batton, Francis
Bauer, Conrad, glass grinder
Beach, Charles, glass mixer
Beacroft/Becroft/Becrost, George, emp gw
Beatty, Duke, glass packer
Bedigee/Bediger (Bedigie?), Francis, glass painter
Bell, Thomas, gc
Benard, I./Jo/L (see Bernard)
Bergen/Bergin/Birgen, James D., gw
Bernard, Joseph P., gb
Berry, Andrew, emp MFG
Beslet
Billet, Charles M., ge
Bishney, Alfred
Blackburn, Ino/John M., bkkpr, MFG
Blink/Brink, Peter, gb
Bonin/Bonnin, Horace, gw
Bottner, Peter, gb
Boughton, James A., pattern-maker
Bourne, Joseph, Supt.MFG
Bourneque/Bourneique/Bonique, Adolf/Adolph, gb
Bournique/Burnique, Joseph, gb

Bouvie, Francis, I, gc
Bouvie, Francis, II, gc
Bower/Bowes, John H., gw
Bradshaw, A.
Brady, salesman (?) in Boston
Brandt, engraver who did M.B. Co. catalogue drawings
Brooke/Brookes, William, MFG emp.
Brooks, James T. gb
Brooks, Thomas W., gc, founded Morse & Brooks Glass Co. in Wlfd
Brounlee/Brounley, William W., gc
Buckley, James, gw
Burgess, Ryan
Burke, James, glass mold maker
Burke, John, gc MFG
Burke, William, gb
Burner, Joseph, gw
Buthan, Daniel
Butler, W.O., clerk MFG
Byrne, William E., gc

C

Cady, John, teaser
Callaghan/Callahan/O'Callaghan, James, gc
Cantillen, J.D. (see DeCantillon)
Carmichael, D., gc
Carr, Harry
Car/Carr, John
Carroll, David, gc
Carroll, William, emp MFG
Carter, George, gb, gc
Case, James, engineer
Casey, John, gc
Chase, J. (Josh?), emp MFG
Chase, son #1 of Josh
Chase, son #2 of Josh
Chaselbert, Joseph
Chilo/Chulo
Clark, Thomas B., bookkeeper
Clifford, John, gc

Cluto, W.
Coffin/Coflin (see Coughlin)
Collette, Alfred, gm
Collette, Joseph, gm
Collins, Andrew P., gm
Collins, James., gb
Collins, James A., gb
Collins, Ino/John, pot-maker
Collins, Michael #1, emp MFG
Collins, Michael #2, pot-maker
Collins, Thomas, McCord's servitor
Collum, Charles, gc
Conklin, James, III, emp MFG
Conklin, Joseph
Conklin, Thomas, ge
Connolly, Frank, gc
Connolly, John
Connors, John
Conway, James, Sr., gb
Conway, James, Jr., gc
Cook, J. Henry, gb
Corbett, James, gw
Corey, Frank L., gb
Coughlin, gb (see Cofflin, Coflin)
Cox, Samuel, gc
Cram, ge
Crane, William J.L., clerk, ge
Crowley, Patrick, gb
Crowley, Timothy, president Glass Workers' Union, Meriden
Croucey/Croucy/Crusie/Cruzey/Groussi/Gruzzie, Frank, gc
Curtis, George R., treasurer
Curtis, John

D

Daly, Michael, gc
Davis, Benjamin
Day, James
Daybill, Charles E., gc
Daybill, Robert, gc

Dean, John, ge
DeCantillon, Edward O., gc
DeCantillon, Joseph, gc
Dehm, John (Ignatz), glass grinder
Delaney, Joseph, glass moulder
Delaney, Patrick
Dempsey, James, ge
Dempsey, John B., gc
Deutzkowski, John
Dillon, Edward
Dillon, James
Dillon, John
Dillon, Joseph
Dillon, Michael
Doherty, Neil, gc
Dolan/Dolin, Michael
Donahue, John H., gc
Donovan, Dennis
Dooley, William
Doran/Dorlan, John
Dormen, Sawney
Dormer, Alonzo B., MFG sloar man
Doyle, John, gc, packer
Drear, William, gc
Duff, Michael, ge
Duff, Patrick J., ge
Duffy
Dugan, C.
Dukes, William
Durgan
Dyson, William

E

Eaton, Benjamin
Ecker, Victor, gb
Eggington, Oliver
Elliott, Benjamin, glass packer
Elliott, Joseph
Ellis, Henry, gc

English, C.
English, E.
Eydam, Adolph, foreman

F

Fahy/Fay, Lawrence, gc
Fairbrother, Francis, gc
Farald, Thomas
Farrar, Edwin C.
Farrar, W.B.
Fasy, C. Robert, designer
Fay (see Fahy)
Fields, Henry
Filebroun/Filebrown/Fillebroun/Fillebrown/Phillebrown, Henry S., ge
Fillebrown, Andrew J., gc
Fitzgerald, Thomas, glass polisher
Fitzpatrick, Patrick C., gc
Fitzpatrick, William, gc
Flaherty, John E., ge
Flanagan, James, ge
Flannigan, William, MFG decorator
Flood, J.
Floyd, Samuel, gc
Fogarty/Fogerty, John
Fogarty, Mary, glass worker, Franklin
Fogarty, Thomas (E.? F.?), gb
Fogo/Fogoo, John, gb
Forbes, Daniel, gc/foreman cutting shop
Ford, Patrick, ge
Foskett, C.C., Treasurer
Fountain, James, gc
Fox, J.
Franklin, George, emp MFG/gb
Franklin, Thomas, gw/gb/foreman MFG
Franklin, Thomas J.

G

Gaffney, James
Gaylor, Gilbert, watchman

Geiler/Geisler, Gilbert
Geisler, Peter, gc
Gilhuly, William, gb
Gilmartin, Michael
Goodman, Owen Jr., gc
Goodrich, Philip, gc
Gordon, M.
Gorman, sloar man
Gracey/Gracie, Samuel, glass mixer
Gradwell, William E., ge
Grady, John
Grady, M.
Gragan/Gragon, Michael, gc
Green, Denis
Green, Joseph
Greenan, William P.
Groussi/Groussie/Gruzzie, Frank (see Croucey)
Guesford, Thomas, gc
Guhr, Charles, gb

H

Habson/Hafson, James
Hack, James, gc
Haden, Samuel
Haggaty/Haggerty, James
Haley, John
Hamilton
Hamlin, Horatio H.
Hamlin, Nathaniel
Hamlin, O.
Handel, Philip J., decorator
Hannon, John
Harvey, Albert, MFG mold rm.
Harvey, Francis/Frank, gc
Haschel, Arthur, decorator
Hastings, Ino/John
Hastings, Watson
Hatch, George E. secretary, sales agent
Hatch, Samuel

Hatton, William
Hauthorne/Hawthorne, Henry, gc
Hayden/Haden, Benjamin, gb
Hayden/Haden, J.
Hayden/Haden, Samuel
Helby
Henriett/Henriot/Henriott, Frank X.
Herriam, Miles
Hibbel, (possibly Sibbel)
Hickbottom/Hickenbottom/Hickinbottom, Oliver
Hoar, Michael
Hobson, George H., gb
Hobson, John gm
Hobson/Hopson, William
Holdbrook, John
Hollingsworth/Hollinsworth, John, gb
Hollingsworth, Thomas
Hollis, Thomas, gb
Holt, Jesse
Hopson/Hobson, William
Howles, Edward
Hubry, Charles
Hull, R.

I

Irwin, William

J

Johnson, Christopher
Johnson, James, gc
Jougleux, Charles, decorator
Jukes, William

K

Kaeppel, Albert G., decorator
Kalenger, Charles, gc
Kane, James, ge
Kane, Michael
Kane, Valentine

Keekes, W.
Keleher, John
Kellehan
Kelley, John Jr.
Kelly, Michael J.
Kennedy, John, gb
Kenney/Kenny, John
Killinger, Charles, ge
King
Klost, August, glass molder
Koeler/Koler, Frederick
Kraemer, Edward, ge
Krepler, Joseph
Kreysler, Joseph
Kroeber, Emil, glass painter

L

Lacey, George H.J., gc
Lacoursiere/Lacourcisiere/Lacourisiere, Edmund J., gc
Lacoursiere/Lacourcisiere/Lacourisiere, William H., gc
Lake, Charles H., bookkeeper
Lanahan
Lane, John S.
Laubinger, Gustav
Laughlin, Michael
Lawson, J.
Lecher/Leecher/Leicher/Leitcher/Letcher/Litscher, John, gc
Lecher/Leecher/Leicher/Leitcher/Letcher/Litscher, Nicholas
Leighton, Henry B., ge
Leitcher, Phillip
Lennon, Ino/John, molder
Lewis, John, packer
Linnane, Thomas
Litscher, Nicholas (see Lecher)
Litscher, John (see Leecher John)
Litscher, Peter, gb
Long, Andrew
Long, James
Long, Michael

Losson, Charles
Lotz (see Lutz)
Lowney, Jeremiah
Luby
Luhr/Suhn/Suhr, Charles
Lunt
Lutz/Lotz, Caspar/Casper
Lyman, William W., MFG president
Lynn, George H., ge
Lynn, Samuel, ge
Lyons, Robert, gc

M

MacGregor/McGregor, John C.
MacNaham, Eugene
MacNary
Magowan/Megowan, James, gc
Mahary/Manarey/Manary/Mannary/McNary, William J., gc
Malcom [*sic*], Daniel, gc
Malley, Martin, gc
Malley, Matthew, gc
Manary/Mannary, William (see Mahary)
Marsh, George
Martin, Arthur, clerk
Martin, Charles, gc
Martin, D.
Martin, Henry, gc
Martin, William, tumbler gaffer
Mathews/Matthews, Edward T., ge
Mathews/Matthews, James S., gc
Mathews/Matthews, Thomas, gc
Mayer/Meyer, Nicholas
McCabe, Arthur R., gb
McCabe, Charles S., gb
McCabe, Frank
McCabe, John, gb
McCabe, Stephen, gb/foreman
McCairns, John, gc
McCarthy, Daniel

McCarty, J.
McClarty, George, gc
McCormack
McCord, Barney
McCue/McHugh, Ino/John H.
McDaniels, William, glass mixer
McDermott, Michael, gc
McElheney/McElheny/McElhiney/McElhiny, Alexander/Elick
McElheny/McElhiny, John
McEthering
McFarland, Luke, gb
McFarlane, Robert, gc
McGee
McGoldrick
McGowan, James (see Megowan James)
McGregor, John (see MacGregor)
McGuire, Edward, gc
McGuire, Matthew, gc
McGuire, Peter W., ge
McHugh/McCue, John H.
McKay/Mackey, Charles J.
McLaughlin, Michael
McMarron, gc
McNall, George
McNall, Henry
McNally, John, gw in mixing room
McName/McNamee, John H.
McNary (see Mahary)
McNee, John
McNulty, Charles E.
McNulty, George
Megowan/Magowan, James
Meyer/Mayer, Nicholas
Miller, Fred
Miller, Victor
Miner
Monahan, Adam, molder
Monahan, James, gc
Monahan, John, gb

Moneypenny, William, gc
Monroe, Charles F., decorator
Montague, M., servitor
Moore, James H., finisher
Moran, Joseph A.
Morris, John F., gc
Mulcahy, James, gc
Mumblo, J.H., gc
Murray, James J. (proprietor, JJM & Co.)
Murray, Thomas
Myers, James, gb

N

Niland/Nyland, James, gc
Niland/Nyland, Thomas
Nounenmacher, John, mixing room

O

O'Callahan, James (see Callahan/Callaghan James)
O'Connor, gb
Olsausky/Olsensky, Theodore
O'Malley, Martin, gc
O'Neill, Edward
O'Neil/O'Neill, John J.

P

Packwood, John, gc
Pairpoint/Pierpont, Thomas J., designer M.B. Co.
Palmer, Joseph, gc
Parker, Edward F.
Parkinson, John, machinist
Parry/Perry, John, gc
Parry, John, II, ge
Peck, Eugene
Pepper, Edwin G., foreman mold shop
Peters, Henry, shade painter
Phillip/Phillips, John, gc
Pierpoint, William L., blacksmith
Pique, Samuel, gc

Pooley, James
Potts, Joseph H., gc
Potts, William, gc
Pritchard, Thomas E., gc

Q
Quinlan, John J., traveling salesman JJM & Company

R
Raarup, Frederick O., molder
Rahaley/Rahely, John, gc
Rawlins, Hugh (see Rollins)
Ray, Benjamin P., MFG, assistant superintendent
Ray, Joseph D., glass presser/engineer
Ray, Noble A.
Reardon, David, gc
Reardon, Michael
Redding, David, gc
Reiman, Gustave J., decorator
Renier, Joseph, gw
Reynolds, John, gc, AFGWU Local 29 president
Riedel, Christian
Riley, James, gc
Rollins/Rawlins, Hugh, ge
Rosenberger, Joseph, gb
Ryan, James F., gc
Ryan, William

S
Savage, Richard, gc
Scheap, Scheb Frank J., gc
Schlegel, Wenzel, ge
Seymour, Hugh
Shepard, Sheppard, Joseph, gc, gb
Short, Frederick, gc
Shorten, Daniel, gc
Sielke/Sielkey, Frank
Sizel/Sizzell, George
Skidgell Isaac, engineer, boiler inspector

Slaman/Slamon/Slammons, Frank J., gc
Smith, Matthew, gc
Stenger, Peter, gc
Steven, Elias H.?
Stimpson, Charles, machinist
Strech, J.
Suhn, Charles
Sullivan, Albert, gc
Sullivan, Richard, blacksmith
Sullivan, William, gc
Swansey, Peter
Swanson, Bernard
Swindel/Swinden/Swindon, Ernest, gc

T

Talmadge Julius/Junius A., bookkeeper
Tarther, John
Taylor, Nathan, bookkeeper
Thomas, August, gb
Thompson, Frank P., foreman
Toohig, Edward F.
Tredennick, Charles H., pattern maker
Tredennick, William

U

Urlac, H.
Urwin, William, globe maker

V

Varki, John
Voit, Reynold, ge

W

Wagenknecht, Reinhold
Waldron, John
Waldron, Thomas, gc
Walker, Charles H.
Walters, Joseph F., gc
Ware, John, gc

Ware, Thomas, gc
Warren, Timothy, sticker-up
Watts, Henry, gc
Weibel, August
Weollenmezen, O.J.
Westermeyer, Christian, gc
Westermeyer, Jacob
Wetherby, Edwin A.
Whalen, Francis W.
Whalen/Whallen, Patrick
Wilbraham, James H., gc
Wilcox, George, MFG president after Lyman
Wilcox, Horace C., Meriden Britannia Company president
Wilkinson, Sarah, mold cleaner
Wilkinson, Thomas, gb
Wilkinson, William D., gb
Wilson, Frank, gc
Wilson, Joseph, gc
Wilson, W.
Winslow, George
Wolff, H. Adolph
Wood, James, gb
Wood/Woods, William F., gb
Wotsfrik, John
Wypel, Andrew

Y

Yance, Bernard
Yesberger, Conrad

Z

Zissel/Zissell, George H., MFG watchman

Products

A partial list of items produced at the Meriden Flint Glass Company, taken from Joseph Bourne's journals, newspapers and trade magazines:

- Jacob's Ladders for Episcopal Fair
- pipes with stems several feet long for Episcopal Fair
- ruby-plated alumets (match-holders)
- opal alumets
- blue-flashed alumets
- etched flint globes
- Argand globes
- shade with socket for Argand burners
- 5" & 7" squat globes
- blue-plated globes
- ruby-plated globes
- flint globes
- opal globes
- Crown globes
- NYRBR (?) lantern globes
- New York & Fishkill ruby railroad lanterns
- lanterns, half blue and half flint
- bulb chimneys
- mustards

- card dishes, crystal or with acid on opal
- mold-blown round dishes for Wilcox Silver Plate
- painted shades
- Hero shades
- No. 1 hinged Hero shades
- etched flint glass shades
- 8"/9"/10"/12"/14" cone shades
- dome shades
- 2-ring French shades
- 10" "four square" shades
- patented Helio-Ceramic Process Decorated Opal Lamp Shades
- etched shades and globes for Iolani Palace, Hawaii
- flint goblets
- Mitchel goblets (a goblet pattern of plain bowl and drawn stem)
- globes, goblets, flower vases for Mitchell, Vance and Company
- Belle goblets; the term "Belle" may refer to engraved bellflowers, a popular pattern
- Belle ware wines
- Belle saucers
- Belle champagnes
- Belle finger bowls
- Belle clarets

Above, left: Leaf-shaped flint (crystal) card dish in the 1877 Meriden Britannia catalogue. Card dishes were placed on a table in the entry hall and were meant to receive visitors' calling cards. *MHS.*

Above, right: Flint goblets etched with the initials H.C.W. (H.C. Wilcox?), made by the Meriden Flint Glass Company. *MHS.*

- Belle decanters
- low-foot creamers
- ruby-plated creamers
- butter drainers
- aquarian (?) dishes, 15" long, 10" deep and 10" wide in wooden boxes (aquariums?)
- pressed stoppers
- lapidary stoppers
- fluted cruet stoppers blown in mold
- cut plated wine bottles, blue, green & ruby
- opal wine bottles
- flint wines
- cordials
- plated clarets
- champagnes
- green wine glasses
- tumblers
- "Mattoni" bowls on low foot
- high-footed Mattoni bowls
- 8"/9"/10" Mattoni dishes
- pressed flint inks
- pressed opal inks
- ruby-, green- and blue-plated cuspidors
- 4"/5" smoke bells
- 6" smoke bells, blue edge
- thermometer tubes
- barometer tubes
- canary colognes and stoppers
- canary bottles
- opal urinals
- ruby, blue, canary and green hocks
- ruby, blue, canary and green finger bowls

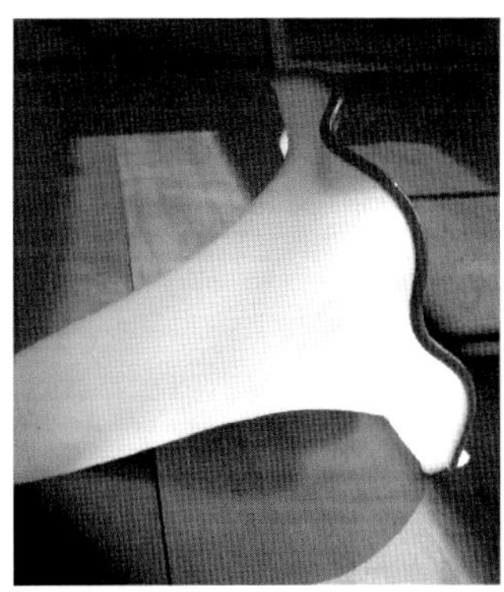

Top and middle: Tumbler. Henry S. Fillebrown, United States, Cambridge, Massachusetts, 1890–99. Colorless glass; blown, engraved H. 9.1cm. Collection of the Corning Museum of Glass, Corning, New York (2010.4.157).

Bottom: Blue-edged smoke bell from a Bradley & Hubbard hanging lamp. Smoke bells hung above oil or kerosene lamps to diffuse the smoke and prevent it from staining the ceiling. *Author's collection.*

- ruby-plated plates
- mucilage jars
- pressed flint pickles
- cut blue flint pickles
- ruby-plated pickles
- blue-plated pickles
- pickle with rim snaps
- reeded pickles of best glass
- oval pickles
- cut blue flint sugars
- low-foot sugars
- puff boxes
- plated celeries
- cut celeries
- 8" trays
- 10" trays, ruby-plated outside
- 12" trays, ruby-plated outside
- 12" flint trays
- engraved 12" trays plated red and ruby
- 18" mold-blown trays
- pillar trays
- 5" nappies (sauce dishes)
- 8" rich cut nappies
- ruby-plated 9" nappies
- 9" peg nappies
- 10" nappies
- 10½" rich cut nappies
- pitchers
- Prince Rupert's Drops
- ruby-flashed-on-white biscuit jars
- 3½" pressed glass boxes
- 2½" opal boxes and covers
- white-on-ruby miniature glass
- flint miniature glass
- glassware blanks for Meriden Silver Plate Company
- decanters
- a 24" x 8" vase
- vases with thread on top
- ruby-on-white plated vases to French patterns

An MFG cut flint "nappie" once owned by Mollie Callahan Nolan, daughter of MFG cutter James Callahan. *MHS.*

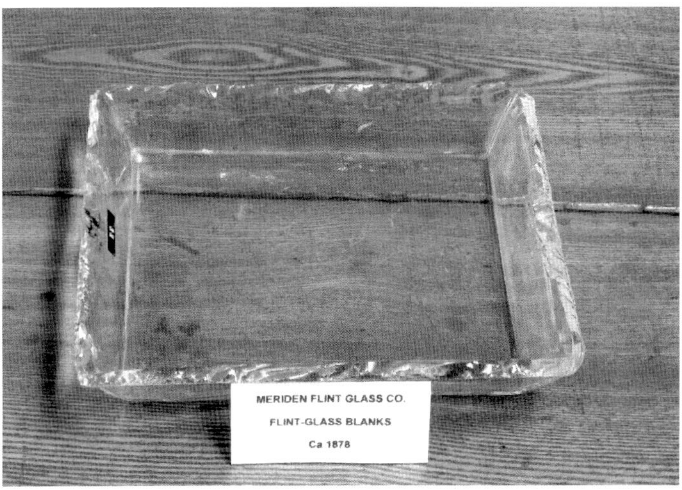

An unfinished square "blank." *MHS.*

- 2-ring flower vases, plated ruby, blue or opal
- sky-blue vases
- large split-ring vase
- cut glass vases held in gilt and French steel standards
- pressed flint vases
- ruby salts
- blue crackled salts
- canary salts
- ruby crackled bottles
- blue crackled bottles
- blue crackled vases
- blue-plated berry dishes

- ruby-plated berry dishes
- comports
- sky-blue jewel cases
- black jewel cases
- 14" dish w/stand
- (infernal) octagon bottles (1878)
- flower pots
- cut glass fruit dishes
- opal fruit bowls
- opal fruit baskets
- mold-blown baskets
- 4-gallon fish globes on foot
- 12" fish globes, no foot
- Hobnail tumblers
- strawberry (diamond shape) saucers
- black glass sconces with white enameling
- cut glass orange bowls

"Star" salt shakers made by the MFG Company for Meriden Britannia Company. The patented insert served to keep the salt from clumping. *MHS.*

Clockwise from top left: A black glass jewel box made for the Meriden Britannia Company; pressed amber peg bowl in a "Tree of Life" pattern set into a Meriden Britannia holder. *Author's collection*; a clear flint bar tumbler. *MHS*; light globe marked "Automatic Lighting Company, Meriden, Conn. USA." James J. Murray's product line included flint globes "for electric light purposes." *Author's collection.*

- ruby-plated Roman punch glasses and stands
- large 2-ring cylinders
- oval dish with foot
- glass water closet bowls
- bar tumblers
- pea green toilet bottles
- toilet boxes
- toilet bottles for Wilcox Silver Plate Company and Meriden Britannia Company

- vinegars
- pepper shakers
- water bottles
- custards
- 9" x 12" bowls
- etched and painted shades and globes for Hogg & Patterson, New York
- Patented Double Globe (1887–88, under J.J. Murray's operation)
- specie jars
- lamp cylinders
- battery jars
- nest eggs

Meriden Flint Glass Company label on the back of the Minton tile. *Author's collection.*

The Meriden Flint Glass Company also decorated Minton tile blanks, as evidenced by this Minton tile trivet bearing a Meriden Flint Glass Company label on the back of the tile, protected over the years by a metal backing. The Winterthur Museum has a brass plant stand with a Minton tile top and an opal glass cylinder bearing a Meriden Flint Glass Company label.

Biographies

Bergen

Glass cutter James D. Bergen was born in Brooklyn, New York, in 1852. He and his first wife, Mary H., had three children, Helen D. (1876), James (1879) and William Tracy Bergen (1880). In 1892, widower Bergen married Nella Reardon, a professional singer. The couple divorced, and James married an actress who was a close friend of his second wife. That marriage lasted only five months, and the couple divorced in 1902.

Flint glass card basket from 1877 Meriden Britannia Company catalogue. *MHS.*

James Bergen sold his interest in the J.D. Bergen Company in about 1909 to Walter Bradley and went to St. Louis to go into the glass cutting business with his son, Tracy. However, poor health forced him to return to his home in Meriden, where he died in 1919.

BOURNE

Joseph Bourne was born in 1826 near Stourbridge, England, where his father was a glassblower. Joseph and his twin brother, Richard, learned the glassblowing trade there also.

Joseph and his wife, Fannie, came to the United States in 1854 with daughters Mary Ann and Henrietta. They settled in Cambridge, Massachusetts, where Joseph worked as a glassmaker for the New England Glass Company and where daughter Emma Frances was born in 1858.

Fannie Moore Bourne outlived her husband by eleven years. Both are buried at Cambridge Cemetery, Cambridge, Massachusetts.

Daughter Mary Ann married attorney George F. Moore. The couple had no children. Three years after Mary Ann's death in 1895, George Moore married her younger sister, Emma. They had no children either. All three are buried in the family plot at Cambridge Cemetery.

Daughter Henrietta married Charles H. Lake in 1874. Their son, Frank, was born in 1876. Frank Bourne Lake attended Cambridge Manual Training School, the American School of Osteopathy and Boston College of Physicians and Surgeons. He and wife Agnes Glenny, both osteopathic surgeons, had no children. In 1918, Frank Lake registered for military service. Agnes was president of the Osteopathic Red Cross Auxiliary, and his mother, Henrietta, was officer in charge of knitting.

FILLEBROWN

Henry Skinner Fillebrown was born in Cambridge, Massachusetts, in 1839. His father, Samuel, was a glass cutter at the New England Glass Company. At about age fifteen, Henry was apprenticed to New England Glass Company engraver Franz Doms and also worked under its master engraver Louis Vaupel.

Henry and his wife, Emily A. Burdakin Fillebrown, had one son, Sylvestus Lincoln Fillebrown.

Henry worked at Hobbs, Brockunier & Company from 1865 to 1870 and at the New England Glass Company before coming to the Meriden Flint Glass Company in 1877. He remained there as an engraver until 1884, when the family "remanded to Boston." Henry worked as a self-employed glass engraver in Medford Hills, Massachusetts. He passed away circa 1910 and is buried at Cambridge Cemetery.

HANDEL

Philip Julius Handel was born in Meriden in 1866. His father died when he was nine, when Philip bought a printer's outfit to help support his family. At age fourteen, he started working at the Meriden Britannia Company, but one month later he became an apprentice at the Meriden Flint Glass Company, where he worked for six months without any pay! At age nineteen, he and co-worker Adolph Eydam entered into the glass decorating business for themselves.

In 1890, Philip married Carolina Sutterlin. She died in 1904 after an illness of two years. No children were born of this marriage.

Philip and Fannie Adelaide Hirschfield married in 1906. In July 1914, the *Hartford Courant* reported that Philip J. Handel had been found in the Handel Company garage with a bullet wound to the temple, a gun at his side and a note marked "Confidential." No reason was known for his apparent suicide at that time.

His widow became the company president after Philip's death and remarried in 1918. Philip's cousin, William, took control of the company, and the firm experienced great growth after World War I. However, it struggled in 1929 when the Great Depression hit. The Handel Company ceased production in 1936. The site of the factory is today occupied by a Super Stop & Shop grocery store, and the Handel home is reputed to be haunted.

LEIGHTON

Henry Barnes Leighton was born in 1837 in Cambridge, Massachusetts. His father, John Hamilton Leighton, was blowing room foreman at the New England Glass Company and son of Thomas Leighton, the New England's first superintendent. Henry was apprenticed to New England Glass Company First Engraver Louis Vaupel in the mid-1850s.

In 1862, Henry enlisted in the Thirty-ninth Infantry Regiment, Massachusetts. After serving during the Civil War, he returned to work in Cambridge as a glass engraver. He came to work at the Meriden Flint Glass Company. He died of unknown causes in 1878, and his widow, Ellen, applied for pension benefits as the widow of a Civil War veteran.

The woman nicknamed "Louie" who Joseph Bourne often visited in Springfield, Massachusetts, may have been Louisa B. Leighton, Henry Leighton's sister. Louisa's first marriage to Peter Vosburgh produced one daughter, Ida B., in 1867. By 1874, Louisa had married John Sidney Smith, who apparently adopted her daughter, Ida, who died at age sixteen. All three are buried in the same plot as Joseph and Fannie Bourne. It is not known what connection Louisa or John Smith had to the Bourne family, but the Cambridge Cemetery plot is in the names of both Smith and Bourne.

McCabe

Stephen McCabe was born in Massachusetts in about 1833. At age twenty, employed as a glassblower in Boston, he married Maria Louisa Byrne. The couple produced five children: Elizabeth, Louisa, John, Charles and Arthur.

Shortly after Civil War hostilities broke out in 1861, Stephen McCabe enlisted in Company C, Massachusetts First Infantry Regiment. In 1862, he was badly wounded in both legs at the Battle of Chantilly and taken as a prisoner of war. He survived the war and returned to Massachusetts to work at the Boston and Sandwich Glass Company. Daughter Elizabeth passed away in 1868.

In 1877, the by now accomplished glassblower came to work with his good friend Joseph Bourne at the Meriden Flint Glass Company, where all three of his sons took up the glassmaking trade.

After his death in 1889, daughter Louisa took over the operation of his businesses for a few Years. By 1900, however, she was living with her brother Arthur and his family in Philadelphia, where he worked as a glassblower.

Monroe

Charles Fabyan Monroe was born in Rhode Island in 1855. While working as a New England Glass Company decorator, he was induced to come to

Meriden, where he remained for the rest of his life. In 1906, he married Emma L. Taft, and the couple lived with her brother and sister in an elegant home at 37 North Parker Avenue.

After closing his business in 1916, Mr. Monroe became foreman at the Edward Miller Company, one of the numerous decorative metalware factories in Meriden. He passed away of typhoid in 1919 and is buried at Swan Point Cemetery, Providence, Rhode Island. The couple had no children. His widow took in boarders at their home for a short while before apparently moving to Plainville to work as a servant.

The site of the C.F. Monroe factory is today an empty lot.

NILAND

Glass cutters James and Thomas Niland learned their trade in Birmingham, England. Their niece, Mollie Callahan Nolan, said in a 1963 newspaper article that their mother "paid half a crown a week for her two sons to learn the glass cutting trade, and they learned it well."[16]

Thomas worked at Boston and Sandwich and the New England Glass Company before coming to the Meriden Flint Glass Company in 1881. After going into the glass cutting business with James D. Bergen, he established Thomas A. Niland & Company in Deep River, Connecticut, with two partners. Within one year, he had sold his interest and returned to work at Bergen's. The Deep River shop continued until 1902.

James Niland served a six-year apprenticeship in Scotland and England before coming to Meriden circa 1882 to work for his brother and James Bergen. In 1891, he married James Bergen's sister, Emma. In 1902, he opened in Meriden his own glass cutting shop, which he operated with his brother, Thomas, for a number of years.

Thomas passed away in 1928. The glass shop scaled back operations after 1935 and closed down in 1959.

PAIRPOINT

Silver designer Thomas J. Pairpoint was born in 1838 in London, England. After serving an apprenticeship in Paris, he worked for the London firm of Lambert & Rawlings. In 1868, he came to Gorham Manufacturing in Providence, Rhode Island. Two years later, he married his wife, Ellen. By

1877, he was the head designer for the Meriden Britannia Company, and he worked there with his brother, Alfred.

In 1880, Thomas Pairpoint and about fifty Meriden Britannia Company workers helped launch the Pairpoint Manufacturing Company in New Bedford, Massachusetts. He severed his connection with that company in 1885 and went into business for himself. He died in Providence in 1902 of complications from diabetes.

Thomas's brother, Alfred, lived in Meriden from 1877 to 1880. He published a book, *Rambles in America*, in which he wrote about his "four pleasant years' sojourn in sociable Meriden."

After leaving Meriden, Alfred Pairpoint was employed as a chaser at Gorham Manufacturing until 1895, when he and daughter Nellie went to live with brother Thomas in Providence, Rhode Island.

Patents

December 7, 1875—Hatch—design for inkstand—#8,831
February 1, 1876—Hatch—"bird" card receiver—#8,946
March 6, 1877—Lyman—match-holder—#9,822
April 3, 1877—Bourne—Hero shade—#189,180
July 31, 1877—Lyman—"Crown globe" glass shade—#10,134
April 16, 1878—Hatch—improvement in tempering glass—#202,544
March 30, 1880—Forbes—glass polishing wheel—#226,054
June 8, 1880—Voit—glass scalloping machine—#228,495
June 7, 1881—Hatch—glass lamp fount—#242,638
June 28, 1881—Hatch—water closet bowl—#243,447
September 13, 1881—Peck—plunger for glass press—#246,972
September 27, 1881—Peck—glass ornamenting machine—#247,515
August 1, 1882—Forbes—glass polishing powder—#262,034
September 19, 1882—Hatch—fruit bowl design—#13,300
September 19, 1882—Hatch—flower vase design—#13,301
October 31, 1882—Talmadge—glass water closet bowl—#266,918
July 24, 1883—Hatch—cobweb cut glass pattern—#14,117
October 2, 1883—Hatch—glass water closet bowl—#286,000

Dealers

Higgins & Seiter

Stevens & Stocking, Cromwell, Connecticut (Hero shades)

Wilcox Silver Plate Company, Meriden, Connecticut

Meriden Silver Plate Company, Meriden, Connecticut

Meriden Britannia Company, Connecticut

Bradley & Hubbard, Meriden, Connecticut

Miller Company, Meriden, Connecticut

Mitchell, Vance and Company, 836 and 838 Broadway, New York

Collamore & Company, 921 Broadway, New York

Steele Brothers, Philadelphia

Hogg & Patterson, New York

R. Hull, Providence, Rhode Island

Taylor Instrument Company, Rochester, New York

G.M. Luther & Company, Providence, Rhode Island

Simpson, Hall & Miller, Wallingford, Connecticut

Plume & Atwood Manufacturing, Connecticut

New York & Fishkill Railroad

Hartford Sanitary Plumbing Company, Connecticut

J.W. Luther, Atterbury, Massachusetts

J.F. Sprague, Providence, Rhode Island

James L. Howard & Company, Hartford, Connecticut

Hutching & Company, Providence, Rhode Island

The Meriden Flint Glass Company maintained a showroom at C.F. Monroe's Art Rooms in Meriden, as well as at 147 Franklin Street, Boston, Massachusetts; 78–80 Murray Street, New York City; and 46 East Fourteenth Street/37 Barclay Street, New York City.

Glossary

agent: general or business manager at the glassworks, or a salesman.

alumet/allumette: a match for lighting lamps, etc.; also, a glass match-holder.

anneal: the process of tempering or cooling glass slowly under a controlled reduction of heat so that it does not crack.

arch: the fire chamber of a glassmaking furnace.

arsenic: a metallic element used as a clarifying agent in the manufacture of glass.

articleing: to "article" refers to a contract binding an individual to work for another for a given period of time.

batch: mixture of raw ingredients ready for melting in the pots.

belle: may refer to an etched pattern of bellflowers.

blank: an undecorated glass item intended for further decorating by cutting or etching.

blast: start of a furnace after summer rebuild.

block fire: reducing furnace heat by 40 percent during temporary suspension of manufacture.

blow over: "an arrangement in blowing glass bottles or jars in molds in which the surplus glass is collected in a chamber above the lip of the vessel with but a thin connecting portion, so that the surplus is readily broken off without danger to the vessel itself" (*Knight's 1880 Mechanical Dictionary*).

bore: hollow center of a thermometer that is filled with mercury.

bye hole (linnet hole): a small hole in the side of a furnace on each side of a pot mouth, connecting flues with chambers, through which flames shoot out about five or six inches; their purpose was to obtain sufficient heat around the pots, the length of the flames indicating the robustness of the furnace heat. Also used for reheating small articles or heating blowpipes.

carry-in boy: an apprentice who carries glass from chair to lear.

cased: glass in which one or more layers are encased in an outer shell of one, two or more layers of glass of other colors.

caster/castor: a vial, cruet or other small glass vessel, used to contain condiments at the table; a metal stand to contain a number of such cruets.

cave: a long air tunnel passing below the furnace, through each end of which air is drawn by the action of the draught of the chimney cone above the furnace to feed the fire.

chair: bench with extended arms in which the master blower sits while working hot glass.

charge: putting the mixed batch materials into the melting pots.

coal gasification: the process used for converting coal to combustible gases by heating the coal in sealed vessels called retorts. The heated coal breaks down into three components: gases, tar and coke.

crisled: a network of cracks in the surface caused by insufficient lime in the batch.

Croton oil: an oil obtained from the plant *Croton tiglium*; a powerful drastic purgative. Applied externally, it produces inflammation of the skin attended with pustular eruption; used as a counterirritant in rheumatism, gout, neuralgia, bronchitis and so on.

"crowning an apprentice": a custom that consists of an apprentice, at the end of his term, contributing a certain sum, and the men with whom he has worked the remainder, toward having a supper and a social event together as a celebration of the occasion.

crucible: pot made of fireclay, in which the raw materials for making glass are melted.

cullet: cleaned broken or waste glass.

Eccles: an inactive volcano in Australia.

flashing: coating one color of glass with a thin layer of another color; frequently, the outside color is cut through to show a pattern against a contrasting background (i.e. ruby-cut-to-clear).

flint glass: originally, glass made from ground flint and litharge, or red lead, producing a very high-quality crystal. The term "flint" later became synonymous with exceptional-quality crystal and continued to be used extensively in advertising and in the names of glass companies to suggest a standard of excellence in their ware. Today, the term flint glass refers to lead crystal.

flux: an alkaline or metallic substance added to a batch to assist in vitrification of silica (vitrify meant to melt into glass or a glasslike substance).

fly: explosion of glass due to inner stress caused by cooling too soon.

founding: the time taken for melting the batch, from the charge until the molten metal is ready to work.

gaffer: master blower, or a shop foreman; sometimes superintendent of several shops.

gasometer: a coal-gas storage tank.

gather: blob of molten glass that clings to the end of a blowpipe, ready for blowing.

gatherer: man who makes the gather for the gaffer (second-ranking man in a shop).

glassmaker's soap: black oxide of manganese that, added to the batch, washes out the yellow or green tinct (tinge) of iron in the sand used in the batch.

glory hole: a small furnace used for the frequent reheatings necessary during the process of blowing and shaping glass.

high color: a pink, blue or violet tinct or tinge noticeable in crystal glass, caused by an excess of decolorizer (i.e. manganese) usually in connection with lead glass.

hobnail: a six-sided, flat-topped cut motif resembling a hobnail of a heavy boot.

Ino: nineteenth-century variant spelling of "John."

lapidary cutting: sharp, angular cutting associated with cutting of diamonds and precious stones.

lear/leer/lehr: an annealing oven.

lime glass: a lower grade of glass ideally made with soda and lime as a flux, suited for common tableware.

linnet hole: see "bye hole."

litharge: lead oxide (poisonous).

low color: green or yellowish-green tinge noticeable in crystal glass, caused by presence of iron in the batch.

lute: a clay compound used to plug any openings in the chimney so as to make it airtight and to protect it from the furnace heat.[17]

manganese: Glassmaker's Soap used to counteract the presence of iron in a batch.

marver: iron slab on which a gather of molten glass is rolled after it is taken from the furnace and before it is given to the gaffer to be blown.

master salt: see "salt cellar."

mattoni: straight-sided oval bowls made in several sizes.

metal: glass, either molten or hard.

miniatures: may refer to one-part molded pieces, which were seamless. One-part molds required a special technique of blowing. The "gather" was blown into the mold and then contracted by means of an inhalation and removed from the mold. It was then blown out to full size, the pattern expanding with the glass. Work of this kind was seamless and miniature in size. ("The Glassmaker's Craft," 1940, http://www.oldandsold.com/articles04/craftsman22.shtml.)

Mitchell: name of a goblet pattern.

moils: the remnants of glass removed from the blow-pipe during glassblowing.

mold: cast-iron encasement into which hot glass is poured or blown to form an object.

mold blown: glass blown by hand into an undecorated paste mold suspended in a tub of water. See "paste mold."

monkey pot: a small melting pot used in making small batches of flint glass. Also known as a "Jockey," as it was sometimes placed on top of another pot.

move: work shift; also the number of items a production crew can make in a four-hour or half-day shift.

nappie: a sauce dish.

offhand: fashioned freehand without the use of either press or mold.

opal/opalescent: opaque or nontransparent, white or colored glass; white opal glass is often called "milk glass."

orguinette: a small, hand-cranked reed organ.

overtaker: apparently a name for a type of flint glass. William Leighton Jr. kept a forty-eight-page Batch Book with numerous receipts (recipes) for different types of flint glass. One of these is for Flint Glass No. 5, which he called "Overtaker."

parison: the glass bubble formed by blowing into a blowpipe.

paste mold: a mold of wood or iron lined with a paste made of beeswax, rosin and other ingredients. This mold is suspended in a tub of water and the vessel is blown hot into it.

pillar-molded glass: glass vessels with mold-blown vertical ribs but no corresponding indentations on the interior. This effect was achieved by partly inflating the gather, allowing it to cool sufficiently to become somewhat rigid and then gathering an outer layer of glass around it. The parison was then further inflated in a ribbed dip mold, which shaped the soft outer layer without affecting the inner layer.

plain: molten glass that has become clear, with no bubbles, and is ready to be worked.

pontil: an iron rod used by glassmakers for manipulating the hot glass; also called ponty/puntel/puntil/punty.

pot: see "crucible."

pot-arch/potarch: a kiln used to heat up a pot to a temperature at which it can be placed into a hot furnace.

pot ring: a fireclay ring that sits on the surface of the metal, or molten glass, preventing the stiff and stringy edge of the metal from mingling with the interior or hotter surface from which glass is gathered for working. Any bubbles or impurities in the molten glass have a tendency to attach themselves to the ring.

pot-shards/pot sherds/potsherds/pot shells: pieces of broken melting pot that are ground and added to the clay used to make new pots.

Prince Rupert's Drops: A small piece of hot glass that, upon being dropped into a tub of ice water, cools quickly into a clear piece of glass shaped like a teardrop. The heavy end can be pounded with a mallet without breaking, but if the tip of the tail is snapped off, the whole piece will shatter into powder. Once used as children's toys.

refractory clay: fireclay.

retort: a sealed vessel or oven in which coal was heated to make coal-gas.

reverberatory furnace: a nineteenth-century flint glass furnace was called a reverberatory furnace as opposed to a blast furnace. The crown, or five-foot-high domed ceiling of the furnace, was supported by firebrick pillars. The heat and flames rising from the eye reverberated from the top of the crown directly down on the pots, and passing between the pots, they were vented out through small openings connecting with flues leading upward to the chimney cone. The upward draught of the chimney drew a steady supply of air into the furnace through each end of the subterranean cave. The strength of the air current depended on the height and capacity of the chimney, as well as how carefully all crevices were luted up.[18]

salt cellar/salt dip: a small glass bowl for holding salt, placed at each dinner table place setting.

seeds: minute bubbles in the metal.

siege: shelf along the interior of the furnace wall on which pots are placed; also, the floor of a glass furnace.

servitor: glassmaker's assistant.

shard/sherd: fragments of broken pottery vessels.

shop: a unit of four or five glass workmen—the gaffer (master blower), the gatherer, servitor, carry-in boy and, if necessary, footmaker; one shop of men usually worked one pot at the furnace, making a particular article.

sloar: the room where the annealed glass was removed from the lears, packed in boxes and sent to the decorating or cutting department.

soda ash: sodium carbonate used as flux to start the melting of a batch; it also furnishes part of the alkali necessary for making glass.

sticker-up boy: the boy who sticks unfinished glass pieces into the glory hole for the frequent reheatings necessary during the process of blowing and shaping glass.

teaser/teasure/tiseur: a worker who regulates the fires in a glassblowing furnace.

Source: http://www.archive.org/details/knightsnewmechan00knigrich.

Notes

CHAPTER 2

1. Kirk J. Nelson, "The Short Biography of Louis Vaupel," *Magazine Antiques*, April 1996.
2. Lura Woodside Watkins, *Cambridge Glass 1818–1888: The Story of the New England Glass Company* (New York: Bramhall House, 1930).
3. *The Encyclopaedia of Chemistry*, vol. 2, *Glass–Zinc* (n.p.: J.R. Lippincott & Company, 1879), 28.
4. John Townsend Trowbridge, *Lawrence's Adventures Among the Ice-cutters, Glass-makers, Coal-miners, Iron-men and Ship-builders* (Boston: Osgood & Company, 1871), 65–69.
5. *Second Annual Report of the Bureau of Labor Statistics of the State of Connecticut, for the Year Ending November 30, 1886* (Hartford, CT: Press of the Case, Lockwood & Brainard Company, 1886), xxxi.

CHAPTER 3

6. Christopher H. Sterling, Phyllis W. Bernt and Martin B.H. Weiss, *Shaping American Telecommunications: A History of Technology, Policy and Economics* (Mahwah, NJ: Lawrence Erlbaum Associates, Inc., 2006), 54.
7. *Glass Collector's Digest*, "American Flint Glass Workers Union (AFGWU)" (August–September 1987): 88–91.
8. George M. Parker, AFGWU International President, "Historical Review—124 Years in the Making."

9. Robert Ellis, *Great Exhibition of the Works of Industry of all Nations, 1851—Official Descriptive and Illustrated Catalogue* (n.p.: Spicer Brothers Publishers, 1851), 703.

CHAPTER 6

10. William Paul Gerhard, *House Drainage and Sanitary Plumbing* (n.p.: D. Van Nostrand Company, 1882), 164.
11. *Second Annual Report of the Bureau of Labor Statistics.*
12. Carroll D. Wright, *Report on the Factory System in the United States* (Washington, D.C.: Government Printing Office, 1884), 67. Based on research done during 1881 and 1882.
13. Jane Shadel Spillman, *The American Cut Glass Industry: T.G. Hawkes and His Competitors* (New York: Antique Collectors Club Ltd., 2006).
14. *Second Annual Report of the Bureau of Labor Statistics.*

CHAPTER 7

15. *Second Annual Report of the Bureau of Labor Statistics.*

BIOGRAPHIES

16. Mollie Callahan Nolan, *Meriden* [CT] *Morning Record*, August 15, 1963.

GLOSSARY

17. Noah Webster, *A Dictionary of the English Language* (Springfield, MA: G. & C. Merriam, 1881).
18. Robert Ellis, *Great Exhibition of the Works of Industry of All Nations, 1851—Official Descriptive and Illustrated Catalogue* (n.p.: Spicer Brothers Publishers, 1851), 703.

About the Author

Diane Tobin attended the University of Hartford and graduated cum laude from Central Connecticut State University in 1989 with a BS in art education, grades pre-K through 12. She taught middle school art and Latin from 2003 to 2011. She is currently pursuing certification to teach elementary and middle school Latin. Over the past five years, she has been invited to speak about the Meriden Flint Glass Company in Connecticut, Massachusetts and New York. She enjoys giving these talks immensely and loves the opportunity to meet so many interesting and interested people.